P9-DMX-874

What the Experts Are Saying

"This is an energetically written landmark book about the virus that causes breast cancer in mice and may very well do the same in women. Ruddy has given us an extraordinary glimpse into the history of this research that will, no doubt, change the world. We should be grateful to her for it is a noble achievement whose time has come."

—Christine Chroner, MD, author of *Waking the Warrior Goddess*

"Dr. Ruddy is a talented and compassionate clinician with an inquisitive mind who is not satisfied by simply providing excellent care to her cancer patients, but who also wants to find a more definitive solution for it. Her passionate quest for a cause has led her to uncover the hidden truths about cancer research, including important discoveries which were not made public and therefore did not impact the natural history of the disease, simply because they conflicted with the mainstream research agenda. Her avid search has important findings regarding the role of viruses in breast cancer, and she provides an informative summary presented with intelligence and passion, along with a balanced presentation of the facts like a true historian. From Loeb and Lanthrop, with their initial observational studies with mice, to Bittner, Andervont, Huebner, and more recently Baker and Holland. The work of these researchers argues for more funding dedicated to investigating the relationship between viruses and cancer. Maybe the availability of more sophisticated diagnostic technologies and increased awareness of the role of the microbiome in human health finally will shine light on this topic and allow us to find answers to the many questions these scientists have raised over the years. Women with breast cancer deserve to know if there will ever be a preventive vaccine. Dr. Ruddy's book involves the reader with its accessible language and passionate narrative, showing how profoundly she cares about her patients. A clinician, a woman, and a scientist looking for clues left over more than a century of research, Dr. Ruddy's book is a must-read for everyone involved in breast cancer."

—Massimo Cristofanilli, MD, internationally renowned authority on inflammatory breast cancer at Robert H. Lurie Cancer Center, Northwestern University, and author of *Nobody Is Listening: Stories of Inflammatory Breast Cancer*

"We are deeply engrossed in the research that I've pursued for twenty years, exploring a virus in human breast cancer that is a 'kissing cousin' of the mouse mammary tumor virus [MMTV], the causative agent for breast cancer in mice. Forty percent of American women who present with breast cancer have this virus in their tumor, but not in the normal tissue of the same breast, which thus excludes genetic inheritance. It is acquired. We are working hard to fulfill the criteria that will prove the virus is indeed the cause of these breast cancers. It keeps me busy and out of trouble."

—James F. Holland, MD, distinguished professor of neoplastic diseases, Icahn School of Medicine, Mount Sinai, New York

"In due course, one could envision, as the ultimate demonstration of a causative link, a vaccination program in unexposed individuals or prepubescent girls that reduced breast cancer incidence."

—John R. Mackey, MD, Cross Cancer Institute, University of Alberta, Alberta, Canada

"We see a need for a consistent protocol for the evaluation of IBC [inflammatory breast cancer] clusters focusing on the laboratory investigation of environmental triggers, primarily infectious agents and chemical carcinogens."

—Paul Levine, MD, The George Washington University School of Public Health and Health Services, Washington, DC

"MMTV is present in human saliva and salivary glands. . . . These findings confirm the presence of MMTV in humans, strongly suggest saliva as route in interhuman infection, and support the hypothesis of a viral origin for human breast carcinoma."

—Generoso Bevilacqua, MD, Pisa University Hospital, Pisa, Italy

"A clear gradient of samples positive for MMTV-like sequences with increasing severity of breast cancer has been demonstrated."

—Caroline Ford, PhD, Prince of Wales Hospital, New South Wales, Australia

"Gestational breast cancers, by definition, are associated with major hormonal changes and hormone responsive elements present in the LTR of HMTV, together suggest a molecular mechanism to explain virus association with hormonally responding tissues."

—Beatriz Pogo, MD, professor of Medicine and Virology, Mount Sinai School of Medicine, New York, New York

"This needs to be in TED talks!"

—Cindy Sullivan, breast cancer survivor and author of the book *Hope, Inc.*

"What we have learned about HIV can be applied to isolate the elusive [human equivalent of] MMTV."

—Luc Montagnier, MD, the Pasteur Institute, Paris, France

"Could a virus be the causative agent for breast cancer, similar to HPV and cervical cancer? In *The End of Breast Cancer: A Virus and the Hope for a Vaccine*, a strong appeal is made for people to recognize a true opportunity to advance the fight against breast cancer. The author, Dr. Kathleen Ruddy, is easily the most persistent and passionate advocate for the scientific world to pay attention to the mammary tumor virus (MTV) as a causative agent for at least some breast cancer. A woman is diagnosed with breast cancer every 20 seconds somewhere in the world, and a woman dies every minute somewhere in the world. It would seem that the real possibility that at least 25 percent of breast cancer might be caused by a virus would have the scientific world excited, and that organizations like the World Health Organization (WHO) would be working at breakneck speed to unravel the mystery of MTV and its association with breast cancer. In this book, Dr. Ruddy has presented the latest evidence that MTV is linked to breast cancer in mice and that MTV has been found in large numbers of persons with breast cancer. And I appeal to the WHO, the National Institutes of Health of the United States, public health agencies around the world, ministers of health, and the scientific community to give consideration to the genuine possibility that the virus is a causative agent of breast cancer in women, at least some of the time. Cancer nongovernmental organizations must not let this opportunity slip. If we can establish the association between breast cancer and MTV, it would go a far way in developing a vaccine and a treatment and in prevention of a significant proportion of breast cancer. For this reason alone this book is a welcome new addition in the fight against breast cancer."

—Leslie Ramsammy, PhD, former minister of health of Guyana (2001–12) and former president of the World Health Assembly (2008)

THE END OF BREAST CANCER

THE END OF BREAST CANCER

A Virus and the Hope for a Vaccine

KATHLEEN T. RUDDY, MD

Skyhorse Publishing

Previously published as *Of Mice and Women: Unraveling the Mystery of the Breast Cancer Virus*

Skyhorse Publishing books may be purchased in bulk at special discounts for sales promotion, corporate gifts, fund-raising, or educational purposes. Special editions can also be created to specifications. For details, contact the Special Sales Department, Skyhorse Publishing, 307 West 36th Street, 11th Floor, New York, NY 10018 or info@skyhorsepublishing.com.

Skyhorse® and Skyhorse Publishing® are registered trademarks of Skyhorse Publishing, Inc.®, a Delaware corporation.

Visit our website at www.skyhorsepublishing.com.

10 9 8 7 6 5 4 3 2 1

Library of Congress Cataloging-in-Publication Data is available on file.

Jacket design by Rain Saukas

Jacket illustrations: iStockphoto

Print ISBN: 978-1-5107-2301-6

Ebook ISBN: 978-1-5107-2302-3

Printed in the United States of America

To my mother, Theresa Garruto Ruddy,
who survived breast cancer and much more.

Contents

Prologue

Lisa

It was the summer of 1995. I had just finished my fellowship at Memorial Sloan Kettering Cancer Center in New York City, and had been recruited by Cancer Treatment Centers of America to create a breast service for the Clara Maass Medical Center, a hospital located in New Jersey. After four years of medical school, five years of surgical residency, and a wonderful year in New York City, I felt that I was ready to take the helm as a breast cancer surgeon and call myself an expert.

One of my first patients, whom I'll call Lisa, was a thirty-four-year-old single mother with three small children. She came to ask my opinion about a lump she had found in her breast six months before. The lump had slowly increased in size during this time, and had become distinctly painful in recent weeks.

I opened my practice right around the time Lisa could no longer convince herself she had nothing to worry about. When she saw an advertisement in the local newspaper for the new breast service at Clara Maass, a hospital close to where she lived, she made an appointment to see me for a second opinion. I will never forget seeing her for the first time. My office was brand new; I think I had four patients scheduled that day. Lisa was sweet and pleasant, her story a little concerning. But when it came time to examine her breast, I was shocked at what I saw. One look was all it took to convince me that Lisa had very advanced breast cancer: the lump in her breast was the size of a lemon, and the lymph nodes under her arm were bulging with tumors. I asked Lisa to explain to me in detail the history of the lump.

When Lisa first discovered the lump, she went right to her gynecologist. The doctor examined her and ordered a mammogram. When Lisa got to the imaging center, the radiologist suggested a breast ultrasound also be done, which it was. But rather than give Lisa a follow-up appointment there and then, the doctor said she would call with the results of the mammogram and they would then decide what, if anything, needed to be done next. But the doctor never called back. And so Lisa assumed what she dearly wanted to believe, that she was fine and had nothing to worry about.

She reiterated that no call back was made and no follow-up appointment was suggested or made. All of this seemed quite odd. I told Lisa to get dressed and meet me in my office, and then I asked my secretary to track down the mammogram and ultrasound reports. It didn't take long for the faxes to arrive, and when they did, I was even more perplexed. The mammogram was perfectly normal, as is often the case in young women. Normally, young women have dense breasts, and this makes other dense masses (like breast cancer) more difficult to see. But the ultrasound was *very suspicious.* It clearly showed the lump in all its nasty glory. It was taller than it was wide, and sported irregular, spiky borders—a sign that it was invading the normal tissues around it. The radiologist concluded that the mass was "suspicious for malignancy" and he recommended a surgical consultation and a biopsy.

Lisa and I sat down and I explained that her mammogram was completely normal although the ultrasound had visualized the lump quite clearly. I told her that because six months had passed, it would be a good idea to repeat the tests to see what, if anything, had changed. I explained that I thought she would need a biopsy but that this would be scheduled only after I had reviewed the new films and discussed the results with her.

I knew Lisa's world was about to shatter into a million hard-to-pick-up pieces, and so, without alarming her, I reassured her that we would get to the bottom of what was going on. I asked my secretary to make the appointments for Lisa's repeat mammogram and breast ultrasound, and asked Lisa to return to my office three days later. I also asked her to bring all her films when she came back so that I could assess how fast her tumor had grown in half a year.

As soon as Lisa left, I called her gynecologist to introduce myself and discuss the case. It was only during our conversation that the gynecologist became aware that the mammogram and ultrasound reports that she had ordered six months before had been filed away in Lisa's chart without her knowledge. She was horrified and immediately stricken with remorse. It was an agonizing few minutes as we pieced together just what had happened. All the while I could hear the doctor rifling through Lisa's chart, frantically seeking a reasonable explanation for the misfiled reports, muttering helplessly about the dreadful mistake that had been made. Unfortunately, this simple clerical error, probably made in the middle of a chaotic day in the office, led to a delay in diagnosis of a breast cancer that might cost Lisa's children their mother. To make matters worse, Lisa was a single mother. Her physically abusive husband had left her, returned to the Philippines, and refused to provide support for her or the children. Lisa worked as a secretary, had no health insurance, and had no immediate family living in the United States.

Lisa was young and otherwise healthy, and that was a great advantage. But she had an aggressive, advanced, and untreated breast cancer, and that was not so good. She was going to need a great deal of treatment to roll back the damage—aggressive treatment sufficient to meet the aggressive nature of her tumor. But more importantly, Lisa and her children were going to need emotional care and financial support.

I performed the biopsy the following week. The pathology report confirmed our fears. Lisa's cancer was aggressive and advanced. But further tests indicated that the tumor was confined to the breast and lymph nodes, and had not spread elsewhere in her body. Lisa's tumor was stage 3, which meant she had a fighting chance. Like so many patients, Lisa embraced her fate with courage and resolve, and pressed on with a grace and fortitude normally reserved for saints.

Due to the advanced nature of her disease, the expert team at the breast service recommended that Lisa's treatment begin with chemotherapy. Hopefully, this would shrink the tumor quickly and allow us to observe how it responded to the drugs. After four months of chemotherapy, Lisa would have surgery and then radiation therapy to the chest wall to help reduce the risk for local recurrence.

The chemotherapy worked well, but not completely. Lisa's tumor shrank to about half its original size. But over the course of four months of treatment, the swollen lymph nodes disappeared entirely. Her surgery went well. There were some cancer cells detected in the mastectomy specimen, but the lymph nodes were negative for tumor. Lisa chose to delay any reconstruction because she would be receiving radiation therapy. She decided to limit surgery as much as possible until her treatments were complete and she felt fully recovered from her ordeal. Lisa remained stoic and brave throughout it all, working on the days that she wasn't actively involved in one form of treatment or another.

Six months later, Lisa slowly began to regain her health, and her hair grew back. Fortunately, the charity care program then in place at the hospital paid for all of Lisa's medical expenses, so she didn't have to labor under that great burden, too. Throughout the remainder of that year and well into the next, Lisa did just fine. I saw her every three months, as is my routine, and always found her cheerful, optimistic, healthy, and grateful for her care.

But then one day, Lisa showed up in my office before her next scheduled visit. I knew something must be wrong. I braced myself as Lisa described the new pain she felt in her ribs. She winced when she raised her arms to unbutton her blouse to show me where it hurt. She could hardly lie back on the exam table because of the pain. I felt her abdomen and found her liver enlarged—filling up with tumor. There was no evidence that the breast cancer had returned under her arm or along her surgical scar, but it was clear the cancer had now set up shop in her bones and abdomen. I told Lisa we would figure out exactly what was wrong and do our best to fix it. She left my office noticeably less upbeat than usual: she knew.

X-rays and scans confirmed that Lisa's breast cancer had spread to her bones, lungs, and liver. When I saw her two days later, I explained to her that radiation therapy could help reduce the pain. And it did. This was a huge relief, for she needed to get back to work. I suggested she apply for disability and assured her it would be granted speedily, which it was. I also told her that I would consult with her oncologist to see what other treatments (in other words, drugs) could be used to help control the new tumors that were springing up like dandelions around her body. The oncologist placed Lisa on a more toxic, but, unfortunately, less effective chemotherapy. The tumors grew more slowly, but didn't go away. A few months later, new tumors started to appear and the old ones began to grow again. Lisa began to lose weight. She had no appetite. Her abdomen and feet began to swell. She was increasingly fatigued. In fact, preparing meals for the children totally exhausted her. At this point, I was seeing her every week. I didn't know how much time was left, but it didn't seem like much.

One day, Lisa showed up in my office unexpectedly. She sat on the edge of the examination table and leaned forward to get closer to where I stood in front of her. With one thin, pale hand she clutched the hospital gown across her sunken chest. With the other hand braced beside her, she kept her body steady and upright. Her head and neck collapsed between her shoulders like a turtle about to retreat inside its shell. She pushed her face into mine and whispered in a barely audible voice that still echoes down the years. Her oncologist had informed her, "There is nothing more I can do." Lisa was terrified, and she began to plead with me, "Isn't there anything *you* can do, Dr. Ruddy? Please, I have three children. I am *all* they have." These words and the recollection of them haunt me to this day.

Medical schools are very good at teaching doctors how to diagnose and treat disease. I have the books and the grades and the awards and the certificates to prove it. Residency training consolidates this immense body of knowledge, augmenting it with clinical application to flesh-and-blood patients. As a medical student, you learn about blood pressure. You learn about what can go wrong and why. You learn the doses and the side effects. When you find yourself with a patient with high or low blood pressure, you ask yourself, "How do I fix the problem? How do I do it without harming the patient, killing the patient, or making the patient hate me for it?" There are thousands of diseases, thousands of drugs, countless procedures, and millions of patients to treat over the course of a medical career. And always the principles remain the same: know as much as possible, do the best you can, and try to improve your skills year by year.

Unfortunately, neither textbooks nor mentors have much to say about what to do when there are no more treatments, no more tests, no more procedures, and no more medications to be given—in short, when there's nothing more that *can* be done. When the last cut is made, the last drug is given, and the last wound is dressed, the last gesture is too often a brief dismissal cloaked in the statement, "There's nothing more that can be done." It's sad but true: when death is imminent, doctors tend to pack their bags and bid farewell. After all, the emphasis in medicine is what and when and how to *treat* disease. The American College of Surgeons' *ACS Surgery* (sixth edition), with its seven editors and 308 contributing authors, confines its discussion of death to (1) how to define it, and (2) how to declare it—neither of which is particularly useful when patients like Lisa still need a great deal of help. These are human beings who may have come to the end of medicine, but they have not come to the end of their lives. More than ever, they need our *care*.

Indeed, there's much that can be done when death approaches. Kindness, tender support, and simply spending time with the patient are critically important and much appreciated. I didn't learn this from books; I learned it from my family, by example. And it's a good thing for my patients that I did.

I told Lisa that I would see what I could do to help. Knowing that she was eager to avail herself of any opportunity to prolong her life, I called her medical oncologist to ask if he knew of any clinical trials in the area that might be appropriate for her. He said, "No, there is nothing suitable." I told Lisa that I would continue to make inquiries at other research institutions in the New York area in case any new investigational drugs became available. Then I gently approached the most delicate subject of all: What plans had she made for the care of her children? Lisa told me that she had "not gotten that far." She was simply unable to face the nightmare of leaving behind three essentially orphaned children. I told her I would help her sort it out and suggested we meet with the hospital social worker right away to get her advice. Then I met with the other members of the breast service to review Lisa's case.

By using a team approach, the members of the breast service were able to draw on their multiple resources and perspectives to find a sensible solution. As Woodrow Wilson once said, "I not only use all the brains I have, but all the brains I can find." In the end, we helped Lisa arrange to have her extended family in the Philippines take custody of the children. Her children were counseled, supported, and prepared for her untimely passing. Although it was a tragedy unfolding, it was not kept secret from the children, thereby avoiding an unexpected shock. Lisa was made comfortable and pain free. The hospital finance department forgave the remainder of her medical bills. Lisa reached out to the pastor of her church who gave her enough money to get her through her final days. By working together, we were able to relieve a great part of Lisa's suffering. She died three weeks later. Her children are all now grown.

This deeply troubling case blew across my young career like a sudden gale at sea, and it steered me in a different direction from where I thought I was headed—from striving for a cure to looking for the cause. Lisa's ordeal—and mine, too—left me wondering as never before, *why did this woman get breast cancer in the first place?*

Like the majority of women with breast cancer, Lisa had no risk factors—none! There was simply nothing to explain her tumor but *chance*; and chance is not much of an explanation, is it? At the time, the causes of breast cancer seemed like a mystery parked on the dark side of the moon. Breast cancer had been made popularly notorious—tied up in endless pink ribbons, bound up in races, dances, stands, and walks, and aligned to any item (including porta potties) painted pink. But the truth of this disease is buried in a mass of grim statistics, along with nearly half the patients, for the overall survival rate for women with breast cancer around the world still hovers around 50 percent. To put a very fine point on this dreadful data, consider this: a woman is diagnosed with breast cancer somewhere every twenty seconds, and another woman dies every minute—each one falling like a leaf from some enormous tree.

***

I first learned of the breast cancer virus in February 2007. It happened quite by accident. Kenneth A. Blank, MD, a radiation oncologist who was a member of the breast service at the Clara Maass Medical Center where I worked, had attended a breast cancer review course in New York City that January and brought the handout to our weekly meeting. He handed it to me with a knowing smile, for he knew I read everything I could get my hands on. I accepted the gift gratefully, with an equally knowing smile.

Ken mentioned there was something in the handout about a breast cancer virus. I thought, *What?*

When I scanned the document during the meeting, I found two PowerPoint slides presented by James F. Holland, MD, at the San Antonio Breast Cancer Symposium in December 2006 about a breast cancer virus—something I had never heard of and never dreamed existed. When I got home that night I read the handout more closely, but because it was just a brief summary and two slides, there wasn't much to learn from it.

The next day I asked Arlene Mangino, my hospital librarian, to pull the entire paper Dr. Holland had presented in San Antonio so that I could study it more thoroughly. I also asked her to pull anything else that he had written on the subject. *Surely,* I thought, *this couldn't have been his first and only paper about the breast cancer virus?* Just before I left the library to return to my office—more as an afterthought than a plan—I asked Arlene to conduct a complete literature search about this virus. I reasoned that other scientists might have written a paper or two on the subject, and I wanted to have as much information as I could reasonably gather as I set out to learn about this most unexpected thing—*a breast cancer virus!*

I was confident that Arlene would probably unearth a handful of articles, and that I'd be able to polish them off in a few hours. I was sure I'd be brought up to speed with the amazing discovery of this breast cancer virus in no time at all. And while I waited for Arlene to get back to me with her literature search and all the articles, I mentioned the virus to everyone I saw over the next few days. No one had heard of it either. Everyone was surprised to discover that such a thing existed.

When Arlene appeared in my office a few days later with a stack of articles two feet high, I was shocked. The first article about the virus was written in 1936. I wondered how there could be so much published material going back decades on a subject as important as this, and *I had never heard of it before.* Then I panicked. Had I fallen asleep during a lecture about this virus? Had I missed reading about it in one of the many breast cancer textbooks that line my shelves at home and work? Had I failed to pay attention during some discussion with my colleagues? I immediately checked my textbooks for references to the virus. Nothing. No citations anywhere. I felt relieved, but not by much. I glanced at the stack of papers piled high on my desk and got to worrying all over again. Never mind the data: What was the history behind this research? Why had it only recently come to the surface at San Antonio? Clearly, I had a lot to learn.

At that point, I thought the best thing to do was to put the articles in chronological order so that I could better follow the logical progression of the research on the breast cancer virus, and I would try to ferret out the political history that had shrouded it later on. I started reading the articles and kept at it for about a month until I finally got to Holland's paper, published in 2006. It was quite a slog, but the more I read, the more intrigued I became. Meanwhile, I was also in the middle of an exceedingly demanding master's program at McGill University in Montreal and was busy traveling the world, creating an international breast service, and preparing my research thesis. I was still running the breast service at Clara Maass, too. Between patients, before bed, on planes, in trains, and during holidays, I read everything I could find about the breast cancer virus. Invariably, I would read one paper and then ask Arlene to pull the references cited in that article. It was like pulling threads on a Persian carpet. There were more and more papers to read, and the more I read, the more I wanted to know.

The pile of articles grew like a beanstalk. The more I read, the more alarmed and passionate I became about this subject. It took me a while to put the pieces together, but eventually it became clear that research on the breast cancer virus had been shunned by mainstream medicine for the better part of forty years. It had been thrown overboard by skeptics and activists in the War on Cancer, its feet sunk in cement by proponents for the "cure."

After reviewing the world's literature on the breast cancer virus, I approached the Susan G. Komen for the Cure Foundation. I felt sure that Komen, the largest breast cancer philanthropy in the world, would be interested in supporting research on the breast cancer virus. We met in Summit, New Jersey. I offered their representatives my credentials and told them about the breast cancer virus. They had never heard of it and seemed as interested as me in helping to get this research back into high gear. The women assured me they would bring the virus to the attention of Komen's executives and scientific advisory board, and they further assured me they would get back to me right away. But I didn't hear a thing. So I called them six months later, and we met again. They were just as warm and welcoming the second time around. Really, they couldn't have been more pleasant or seemed more interested in what I had to say. Again, they promised to be in touch, and once again I never heard a word from them. After another month of waiting, the words often attributed to W. C. Fields came to mind, "If at first you don't succeed, try again. If you fail the second time, give up. There's no point in making a fool of yourself."[1] Unfortunately, I felt I would have to give up on Komen. I had no intention of giving up on the virus, however. (*Note:* From time to time, the Susan G. Komen for the Cure Foundation has provided some modest grant support for research on the breast cancer virus. But their representatives in Summit were not aware of this when I approached them in 2008.)

A few months later, as I was finishing my master's program at McGill and wondering what I would do with my expertise as a breast cancer surgeon and my new degree, I began to reflect on what was the next best thing to do. I thought, *I'm a breast cancer expert. I'm an international health care leader. Why don't I create a foundation whose mission is to understand the causes of breast cancer and to use that knowledge to prevent the disease? Yes, that's exactly what I'm going to do. And while I'm at it, I'm going to see what I can do to bring more attention to the subject of this breast cancer virus.*

After arriving at this decision, I called my attorney, Jeff Pompeo, and told him I wanted to create a breast cancer foundation focused on prevention, one that could bridge the gap between the cure and the causes of the disease. In April 2008, we filed the necessary papers to obtain a 501(c)(3) nonprofit status from the IRS for the Breast Health & Healing Foundation. Its official mission is as follows: "To discover the specific causes of breast cancer and use that knowledge to prevent the disease." The paperwork was the easy part. The hard part—something quite new to my repertoire—was to figure out a way to raise awareness and money for research on the breast cancer virus.

According to my back-of-the-envelope calculation, less than $100,000 a year (of the hundreds of millions of dollars spent annually on breast cancer research) was given to scientists who were working to identify the human breast cancer virus. *How can I change this sad arithmetic?* I wondered. Well, I didn't know anything about raising money, but knew how to write—it had been my first love in college. So I created a blog and began to write about the primary prevention of breast cancer, and I emphasized the importance of supporting research on the breast cancer virus.

That certainly got some attention. My blog has since won two awards, and if you do a Google search on the breast cancer virus (a.k.a. the "pink virus") you will find my name and website mentioned. But that wasn't enough. More women needed to know, not just those who poked around on the Internet and found their way to me. I held two summits on Capitol Hill about the virus. I cosponsored another summit in New York City with the help of the Harvard School of Public Health and Loreen Arbus of the Arbus Foundation. I made this project my "Commitment to Action" for the Clinton Global Initiative. I made a public service announcement about the virus, a short documentary film (*It's Time to Answer the Question)* that was nominated Best Film of the Year 2010 by Rethink Breast Cancer, a Canadian breast cancer foundation. And then I thought, *I must write a book about the virus. Women need to know!*

My goal in this book is a simple one: tell a good story that will draw enough attention to this research that scientists who are working on the breast cancer virus can get the money they need to finish their work. You see, I believe that answering the question "Does a virus cause breast cancer in women?" has the potential to change the life of every woman on the planet. If scientists were to prove that this breast cancer virus is responsible for a large portion of human breast cancer, it would change our entire approach to the disease. The "Pure Cure—Prevention"[2] could begin in earnest. Scientists could screen women to identify those who are infected with the virus and are therefore at an increased risk for breast cancer, just as we now screen women for the human papillomavirus (HPV) that causes cervical cancer. Scientists could design interventions, like antiviral medications, to reduce the risk for breast cancer in women who are infected, just as patients who are HIV positive are given medication that prevents them from developing full-blown AIDS. Doctors could identify breast cancer survivors whose tumors were caused by the virus and develop specific therapies to reduce their risk for recurrence and death. But we can't do any of this until the investigations are complete and the proof is in.

Yes, we can continue to fight the War on Cancer that President Richard Nixon launched in 1971 in the same fashion and with the same entrenched rules of engagement that have prevailed for the last forty years: diagnose, treat, and work hard for a cure. Patients have been and will continue to be thankful for everything doctors can do to help them in their fight, and they will be forever grateful when the treatments they subject themselves to deliver a durable cure. Our great scientists can continue to make incremental and occasionally impressive advances in the way cancer is diagnosed and treated, even though the unremitting rise in the number of new cases will soon be a burden that no country, not even the wealthiest, can afford.

All of these converging activities have made a huge difference in the lives of patients over the past four decades, and, no doubt, they will continue to do so. But it's simply not enough. If a large portion of breast cancer is preventable because it is caused by a virus, then wouldn't we be better served if the leaders of breast cancer philanthropies and directors of the relevant government agencies moved more definitively toward the "Pure Cure" (i.e., primary prevention of breast cancer) by taking deliberate, forceful, and committed steps to provide the money necessary to complete this research? I can't help thinking so. Surely, Lisa and millions of other women like her would have given anything to have been spared their early, tormented deaths, if not for themselves, then certainly for the sake of their children.

Chapter 1

The Immortal Renegade

On a beautiful, warm September morning, two women, each thirty years old, wake up happy and well. One woman jumps out of bed, takes a shower, and while soaping up, finds a small lump in her breast. The other woman wakes up early and begins her day by having sex with her husband. He finds a similar-sized lump in her breast, briefly recoils, and then points it out to her. She feels it, too. And thus, three more people who have long assumed that breast cancer is something that happens to other people immediately begin to flounder in an undertow of dread.

Both women live in the same town, and both happen to share the same gynecologist—we'll call her Dr. Jean. The women keep themselves busy doing household chores until Dr. Jean's office opens. At precisely 8:00 a.m., they grab their smartphones and dial Dr. Jean's office. The receptionist who answers is kind and accommodating, even though she has just walked in the door herself, has five other calls on hold, and is texting her son to make sure he got off to school with his lunch that day. The receptionist has been with Dr. Jean for twenty years, she knows both of these women well, and she is more than familiar with the high anxiety that is now washing over them. She squeezes them into Dr. Jean's schedule for the next day and promises to call right away if there is a cancellation that would allow them to come in sooner. The women are grateful, but not in the least relieved. They spend the rest of the day, and most of the night, in a fog disguised as a flurry of hyperactivity: all the laundry is done, groceries are purchased to last two weeks, the bills are all paid and the accounts brought up to date, clothes are taken to the cleaners, stamps are purchased at the post office, and magazines are sorted and discarded.

When the women arrive for their appointments the following morning, they are quick to tell the nurse that they've "never had anything like this before," as if this statement is a sound argument for why it shouldn't be there now. They say they are "in perfect health," but they know their "perfect health" is no longer the case the way it was two days ago. They add, in a tone bordering on defiant, "No one in my family has breast cancer," as if that should exempt them from such a fate. Sadly, approximately 70 percent of women diagnosed with breast cancer in the United States have no family history; but most women don't know that. They think that breast cancer runs in families (which it does) and that families without a history of breast cancer remain outside the danger zone (which they do not).

The nurse understands what they're trying to say: "This can't be happening to *me*." She nods and jots down other information, particularly the details of exactly when and how they discovered the lump. She patiently nudges them in the direction of the important details she knows Dr. Jean will be interested in: pain, skin changes, nipple discharge, last menstrual period, and method of birth control. The women say that the lumps are not painful. There's no history of ovarian or colon cancer in the family, which, if present, would indicate an increased risk for breast cancer. The women started their periods when they were twelve years old and have had regular menstrual cycles ever since. They have never used birth control pills. They've never been exposed to radiation therapy (another important risk factor for breast cancer). They have had one pregnancy each, naturally conceived, and delivered a healthy child (male) at full term. Their boys are five years old and have just started kindergarten. They show the nurse photos on their smartphones of the children on their first days of school. One of the boys thought his mother would take a picture of him every morning before school, and was surprised to learn she wouldn't. The nurse asks the women what their boys want to be for Halloween. The answer is the same: Power Rangers. Both women nursed their babies for several months and had no problems doing so.

The nurse continues with her inquiry: The women don't smoke and never have. The women both drink a glass or two of wine per week, mostly white. The nurse asks them if they exercise. Yes, one woman attends the Bar Method twice a week; the other woman walks three miles every morning after she drops her son off at school. Both women are stay-at-home moms who look forward to going back to work someday, though not yet. The nurse asks them about contraception. One woman is trying to get pregnant: she's the one whose husband found the lump the previous morning. The other woman hasn't decided if and when she'll have another baby. For the time being, her husband is using condoms.

The women are both white, college educated, and come from European stock—a mixture of German, Irish, and English heritage. Other than the fact that these women enjoy an occasional glass of wine (which increases the risk for breast cancer only slightly if consumed at the rate of one to two glasses per week), there's not a single significant risk factor for breast cancer to be found anywhere—except that they are women and breast cancer is the most common female malignancy in the United States and every other country around the world.

The nurse records their weight and takes their vital signs. Yes, they seem to be in perfect health. She records this information using electronic medical record (EMR) software and then leaves the room saying, "Please take off your top and bra, and put on this paper gown so that it is open in the *front.* The doctor will be right in." Both women know what that means—another 15 minutes, at least. Each woman sits on the edge of the examination table, wrapped in a paper camisole the size of a pup tent, and tries to remain calm, but cannot. The EMR computer screen stares back at them from across the room, their personal data splayed out with the harshness of a roadside billboard. The real story—the bewildering ordeal these two women and thousands more that day must endure—cannot be captured or recorded in data entries on monitors or tablets. Their stories need to be told in words, but words fail them now. It's as if their minds are fogged in with worry, so nothing now seems clear.

The women perch like birds on the edge of the examination table, watchful and alert. They hear voices from staff, patients, and other doctors moving in brisk steps back and forth along the linoleum corridor. They listen to the sounds of doors opening and closing in asynchronous disharmony. Occasionally there is a raised voice or a loud shout down the hallway meant to get the attention of a doctor who's needed on the phone. Then, at last, there is the quiet rap of knuckles on the door and in walks not Dr. Jean, but someone else. Someone new.

Dr. Jean has hired a physician assistant (PA). She introduces herself as Mrs. Holt and then reviews the medical history, with particular attention given to the history of the lump. She examines the women's breasts thoroughly and gently. The breasts are symmetrical: nothing is out of shape or distorted in any way, either when the patients raise their arms or when their arms are at their sides. There are no skin or nipple changes noted. Each woman's lump is easily felt and measures 1.0 centimeter across—about a quarter inch, the size of a hazelnut. The lumps are firm. One lump is fixed to the surrounding breast tissue—that is, it seems tethered to the breast itself. The other lump moves freely, as if it were a lost marble that rolled into the breast one day. There is no evidence of infection in the breast: no redness, no warmth, no skin dimpling. The women don't feel any pain when Mrs. Holt presses down on the lump or the surrounding tissue. There is no enlargement of the regional lymph nodes—those under the arm or the ones that are located just above the collarbone. There is no abnormal nipple discharge. And so, with the exception of the lump, everything seems normal. Mrs. Holt reassures each patient, explaining that Dr. Jean will see them on the next visit, and gives them prescriptions for a mammogram and a breast ultrasound. She asks each woman to return to the office to be seen by Dr. Jean the following week.

Both women are given prescriptions for "diagnostic" rather than "screening" mammograms. There's a big difference between the two types of tests. Screening mammograms are ordered when there are no breast abnormalities, the patient is over the age of forty (or because of family history is deemed to be at high risk under the age of forty), and the objective is to look for radiographic hints of a clinically occult breast cancer (that is, a cancer that is too small to be felt on physical examination). Diagnostic mammograms are ordered when something about the breast is found to be abnormal. They usually involve additional views—images taken from different angles—if needed. Because these two women definitely have something wrong—lumps in their breasts—they need diagnostic mammograms, not screening mammograms. Mrs. Holt, the PA, also orders breast ultrasounds. Breast tissue in young women is normally quite dense. A breast mass such as a lump, which may itself be dense, is not easily seen on a mammogram against a background of breast tissue that is already dense. A breast ultrasound, on the other hand, is far more likely to visualize a lump, even one surrounded by dense breast tissue. Breast ultrasounds are particularly good at distinguishing a solid mass (no fluid present) from a cyst (fluid present). On occasion, a breast mass may consist of both solid and cystic components, in which case it is referred to as a *complex mass.*

Each woman is able to schedule her mammogram and breast ultrasound that week. Both mammograms are completely normal. The breast ultrasounds, however, clearly show the lumps; and both are solid. Each measures a little less than one centimeter, in keeping with the size of the lump measured in the doctor's office.

The women return for their follow-up visits the next week, each arriving early, and with a list of questions she's collected from friends, family, and the Internet. The women are put into examining rooms again and told that the doctor "will be right in." Twenty minutes later, Dr. Jean comes in. Her calm and pleasant manner is reassuring.

She reviews the EMR, examines the patient herself, and discusses the results of the tests. She explains that the next best step is to perform a biopsy. She tells them that there are two ways to do a biopsy. With a needle, in which a small sample of the cells from the lump will be removed and examined under the microscope, or in the operating room, in which case the lump will be removed entirely and then examined under the microscope. Both women say emphatically, "I want it out." Dr. Jean refers them to a breast surgeon in town—Dr. Glenn, a woman who, fortunately, has done a fellowship and has been in practice for ten years. Dr. Jean says she is very good.

Back at the front desk, the women ask the receptionist to call Dr. Glenn's office and arrange for the next available appointment. Yes, the breast surgeon can see them the following week. Another week of waiting. Solid lumps in their breasts, and another week of waiting. It's excruciating, the uncertainty. The worry is a crushing burden. Pink, the color of sweet femininity, is not the color that resonates with them during this anxious travail. Perhaps gray like fog would be a better choice.

Dr. Glenn repeats the history and physical examination all over again. She reviews the mammogram and ultrasound, puts them up on her view box, and points out to each patient what she sees and what it means. As requested, Dr. Glenn removes both lumps in the operating room the following week. (Another week of waiting!) The same pathologist, a specialist in diseases of the breast, reviews the slides for both patients. Two sleepless, nail-biting days later the women return to Dr. Glenn's office to have their wounds examined, their dressings changed, and to learn the results of the pathology report.

The husbands, who were not able to come for the initial consultation, made it to the hospital when their wives had their biopsies and now accompany their wives to Dr. Glenn's office for the postoperative visit. The verdict will not be rendered to just two women; it will be handed down to two families and many friends. The husbands lean against the wall in the examination room, beneath a row of handsomely framed medical certificates. The men are mute and awkward.

Dr. Glenn begins by smiling encouragingly at the first patient and her husband, and gently lifts the bandage to check the wound. The incision is healing nicely, and there are no signs of infection.

Dr. Glenn then tells the first woman that the results of her biopsy were just fine. "Your tumor is benign," she says. "It's completely gone and it won't come back." Everyone smiles with relief. The patient takes her first full breath in weeks and exhales a world of worry. The world is radiant again, and she's glad. She can walk out of the doctor's office and resume her life. Her husband looks even more relieved than she. They reach for each other and exchange a kiss of sweet affection and deep relief.

Later that day, Dr. Glenn has the unsettling task of going through the same post-op motions, but with a different twist. Instead of glad tidings, Dr. Glenn has to tell the second woman and her husband that she has cancer. Dr. Glenn puts it to her in a quiet, gentle way: "When the pathologist looked down the microscope, he saw cancer cells." These words are crafted carefully, not to minimize the diagnosis, but to place the emphasis on the *cells* in order to create distance, a neutral vantage point from which the patient can delve deeper into her new reality without falling into the abyss of the diagnosis of cancer. Dr. Glenn's wordsmithery is designed to soften a horrible blow and allow the devastating information to penetrate slowly into the patient's psyche like a sponge absorbing water rather than like a sharp rock landing on her head. The husband, who stands by casually with one shoulder resting against the wall as he checks his phone for texts and emails, goes from mute to motionless when he hears the news. Dr. Glenn remains calm, standing easily by the patient's side as if she had nothing else in the world to do other than remain there indefinitely to provide comfort and support. Neither the patient nor her husband has much to say. Dr. Glenn says she will meet them in her office across the hall so they can sit and talk about the diagnosis further. This short interval alone, as the patient gets dressed and her husband helps her gather her things, gives the couple time to gather their thoughts, get their bearings, regain their strength, and search their minds for the most important questions that immediately come to mind.

***

What is cancer, exactly? What did the pathologist see when he peered down his microscope that told him: "This patient's lump is fine—it's benign" or "This patient's lump is not—it's cancer"? What, exactly, are the distinguishing characteristics that differentiate benign tumors from those that are malignant? More importantly, what are the consequences that follow from these all-important differences?

Let me begin by addressing two common misconceptions about benign versus malignant tumors: one having to do with pain and the other having to do with the rate of growth. First, while it's true that malignant tumors tend to grow more quickly than benign tumors, this is not *always* the case. Several years ago, a doctor sent me a patient, an elderly woman, who had a small mass on her mammogram that he and the radiologist had been "keeping an eye on" for five years. The referring doctor, an internist, told me that the patient's breast exam had been perfectly normal on every occasion. The radiologist thought the spot on the mammogram looked benign and suggested a "watch and see" approach. As recommended, the mammogram was repeated six months later. The spot remained the same and continued to look entirely benign. The patient's mammogram was repeated faithfully every year for the next four years. And for the next four years, the spot didn't change a bit.

One day five years later, the patient felt a lump in her breast. She called her doctor and was seen immediately. The doctor examined the patient, recorded the size of the lump (three centimeters, about the size of a walnut), ordered a diagnostic mammogram, and sent the patient to see me right away. When I examined the patient, it seemed that her lump was located in the general vicinity of the spot on her mammogram, the one that had remained stable and benign in appearance for so many years. The diagnostic mammogram revealed that the spot had grown and no longer looked benign, it looked like cancer. A big cancer. The lymph nodes under her arm were enlarged and filled with tumor, too. Naturally, everyone was upset—the doctor, the patient, and the radiologist. *Upset* is probably too weak a word; *crazed* would be more accurate.

I performed a needle biopsy and, yes, the lump was a cancer. Indeed, the "spot" on her mammogram, though stable and seemingly benign for many years, had been a cancer all along. This malignant tumor, however, was a very slow grower. Having appeared out of nowhere one year, and giving every impression that it was benign, it went into hibernation for the next five. Then suddenly it sprang to life.

Yes, of course, this is an unusual case. Most cancers start to grow and continue to grow. In fact, they tend to grow faster over time. But growth, or lack of it, is not the primary criterion for judging the nature of a lump or a spot on a mammogram. Cancers may grow quickly or they may grow slowly. On occasion, they may appear out of the blue and not grow at all for quite some time. The take-home lesson is this: anything that's new on physical examination or on a mammogram should be evaluated fully, whether or not it's growing fast.

It's also worth pointing out that pain is no more accurate than growth rate for determining whether a lump is benign or malignant. In fact, approximately 15 percent of women with breast cancer report pain as a symptom. Anything that's growing rapidly in the body will tend to push up against the surrounding tissue, and often this produces pain—in the same way that getting squeezed into a crowded subway will make your shoulders ache. There's only *one way* to determine if a lump is benign or if it is malignant, and that's to do a biopsy. Let the actuaries deal with the probabilities. If you have a lump or a new spot on a mammogram, you'll want to know, for sure, what it is.

How, exactly, does a pathologist decide whether a tumor is benign or malignant? Let me take you through a simplified version of the process. Initially, the pathologist examines pieces of the tumor cut into very thin slices and views them under the microscope. She looks at the general structure of the cells, scanning through the slides to check for any architectural changes *within* the cells or *between* the cells that might give a hint that something abnormal is going on. Changes in the shape, position, and configuration of the cells are the first clues that distinguish benign from malignant tumors. These architectural differences are similar to the features we use to tell one type of house from another: a Cape Cod, a ranch, a Victorian, a colonial, and so on. Each home is different, yes, and these differences are easily recognized as variations in their architectural features. A Cape Cod has two levels, but the second story is built into the roof; a ranch is a single-story home whose rooms may run off in any direction; a Victorian typically looks like a Martha Stewart wedding cake built for habitation, and so on. And, of course, there are also homes that are a mix of styles and that are more difficult to characterize. For example, a California bungalow might look like a cottage, but with a few Victorian flourishes. Likewise, the cells that comprise a tumor, and the relationship of those cells to the normal cells around them, produce recognizable architectural features that can be used to distinguish a benign from malignant tumor. As with our California bungalow—or is it a cottage?—sometimes it can be hard to tell exactly what is going on. It's good to know that in most cases a cancer looks like cancer. (If there is ever a question, get a second opinion. Don't hesitate to take advantage of the fact that there's always more than one pathologist at your disposal.)

There are three characteristics of cancer cells that speak to their murderous enterprise: they *invade*, *metastasize*, and *regenerate immortally*. Let's take each one separately and consider it fully, for each characteristic deserves a thorough explanation so you can better understand what you or another woman is up against when the pathologist says, "This tumor is malignant," and the doctor says, "You've got cancer."

Hallmark #1: Cancer cells *invade*. Like Hitler's army, cancer cells cross borders and move to places where they do not belong. Under normal circumstances, borders (in other words, tissue boundaries) are established throughout the body to function like property lines in the outside world. Consider the liver, an organ with a strong perimeter called a *capsule* that functions like the national boundary between the United States and Canada. Under normal circumstances, liver cells do not grow beyond their liver capsule. When they do grow, they know when and where to stop. But liver cells that have become malignant encroach upon new territory, just like Hitler's soldiers invading Poland. A malignant tumor of, say, the lung can eat its way into the esophagus, the windpipe, the ribs, the spine, and even the heart—eventually strangling the patient from the inside out. Benign tumors may grow exuberantly, even obnoxiously, but they never grow invasively. At most, their growth may press up against the borders that separate them from their neighbors, but they never jump the fences. Cancer cells, on the other hand, will stop at nothing; no obstacles get in their way. It's no wonder that cancers were referred to in the eighteenth century as *morbid growths*. Morbid, malignant, malicious, and invasive growths, indeed.

So, when the pathologist looks down her microscope and sees architecturally bizarre cells crossing borders and growing into places where they do not belong, this is a sign of *invasion*. It is the first hallmark of cancer. It tells her that the cells are malignant and that the patient has cancer.

Hallmark #2: Cancer cells *metastasize*. Malignant cells are not just on the move locally and invasively; they're on to distant lands. In short, cancer cells deploy. They invade existing blood vessels and use them like private access lanes to circulate throughout the body, exiting at favorable sites like the liver, brain, and bone to establish new cancer colonies in remote locations. Cancer cells also invade the lymphatic channels that run alongside blood vessels. Once this happens, cancer cells will frequently set up shop in the lymph nodes that drain the lymphatic vessels. In the case of breast cancer, cancer cells will invade the lymphatic channels and end up most frequently in the lymph nodes located under the arm (in other words, the axillary lymph nodes). Breast cancer cells can also find their way into the lymphatic vessels that run along the breastbone, and then establish new cancer colonies in the internal mammary nodes located there. Or they can move into the lymphatic vessels located above the collarbone and begin to grow in the supraclavicular lymph nodes located there.

Malignant cells take advantage of these two circulatory highways—the blood vessels and the lymphatic channels—to transport cancer cells around the body to distant organs. The lungs, bones, liver, and brain are favorite destinations, being organs where cancer cells have access to abundant oxygen and nutrients—giving them all they need to establish new colonies that will grow, invade, and redeploy.

Worse, metastatic deposits are often lethal. You can live without your breasts, but you can't live without your liver, lungs, or brain. Metastatic cancer colonies that begin to grow in vital organs can be exceedingly difficult to contain. Local treatments that target the primary tumor (for example, surgery and radiation therapy) are highly effective in getting rid of the original, primary cancer. But chasing cancer cells around the body with the intent of killing all of them can turn into a frustrating, frenzied circus. Chemotherapy drugs can help, but they seldom eradicate all the cancer cells once the tumor has metastasized. Cancer cells are notorious for finding ways to adapt to chemotherapy (they have very clever genes), which renders them increasingly resistant to one drug and then another. Sooner or later, doctors run out of drugs to throw at the resistant and resilient tumors growing everywhere.

Hallmark #3: Cancer cells *regenerate immortally*. While invasion and metastasis are cancer's offensive strategies, regenerative immortality is its best defense. By comparison, normal cells are slated to stop regenerating and eventually die—it's part of nature's master plan. Death is the price life pays for progress. It's the underlying cost that bankrolls evolution. Consider for a moment if the first living organisms had lived forever, regenerating immortally and piling up endlessly. There simply would have been no room for us, or anything else. Evolution would have stalled in a swamp of single cells—bacteria and the like. The earth would be little more than scum, not the complex planet we've come to appreciate. Evolution requires a self-cleansing canvas upon which change and adaptation can take place. That means that every living thing must one day stop reproducing and eventually pass away.

Cancer cells apparently did not read the memo about death and dying. If they did, they've found an ingenious way to completely disregard it. Cancer cells will continue to regenerate and reproduce until they are actually killed. As a result, cancer cells can outreproduce and outlive every other cell in the body. Now, that's a survival skill that's hard to beat.

Here's an example of the power of cancer's regenerative immortality. In her book *The Immortal Life of Henrietta Lacks* (see Bibliography), Rebecca Skloot tells the story of a young woman, Henrietta Lacks, who died of cervical cancer in 1951. Lacks's cancer cells were the first to be successfully grown in a test tube. Once they started growing in the laboratory, they were found, to everyone's surprise, to be *immortal*. Lacks's cervical cancer cells, taken from her body the year before she died, have replicated *endlessly*. This was the first we knew of cancer's regenerative immortality. Today, there are more of Henrietta's tumor cells alive than when she walked the Earth—fifty million tons of them, at last count. The "HeLa" cells, as they are now called, have been invaluable to scientific research. They were used to develop the polio vaccine. They helped us understand how cancers form and grow. They were launched into orbit around the moon. Who knows where they will land next? One thing we know for sure: HeLa cells, like all cancer cells, will grow until they're killed.

You might wonder where cancer cells come from in the first place. After decades of looking for an answer to this question, scientists have come to believe that cancer cells evolve from normal cells whose genes have become either deformed or dysfunctional. There's evidence to suggest that cancer cells originate in a single normal cell that, for one reason or another, is led slowly and sequentially down the genetic road to treachery. A series of mutations in the genes of a normal cell—inherited, acquired, or both—corrupt the DNA of the cell, leading to a cascade of subsequent genetic alterations, like a series of falling dominoes, that concludes either with the death of the cell (the most common outcome) or the creation of a malignant mutant that regenerates endlessly to form a cancerous tumor whose immortal cells are capable of invasion and metastasis. Scientists have discovered that this is not an entirely random process: there is an order to this genetic cascade that proceeds from an entirely normal cell to an immortal, malignant renegade. Once the cancer cascade is set in motion—once those dominoes begin to fall—it takes fewer and fewer changes in the DNA to complete the transformation of a normal cell into a cancer terrorist.

More recently, scientists have found evidence to suggest that cancer cells originate from a particular type of normal cell called a stem cell. Stem cells are precursor cells, undifferentiated cells with the power to transform themselves into any type of cell in the body. When you think about it, this is a pretty terrific resource—like having bolts of cloth on hand that can be used to make any article of clothing that might be needed, and at a moment's notice. The body keeps a supply of stem cells on reserve in every tissue in the body that can be "called up" when new cells are needed, such as when there is an injury or when it's time to grow. As an example, during pregnancy, breast stem cells produce new breast cells in preparation for milk production. As many women will readily admit, swollen breasts are often the first sign (after a missed period) that pregnancy is under way. Once breastfeeding ceases, the newly added breast cells, which are no longer needed, are thanked and retired.

Scientists believe that stem cells are more sensitive to carcinogens than fully differentiated cells. As a consequence, stem cells are more vulnerable to genetic mutations than mature and designated cells. Thus, stem cells may be the primary reservoir from which cancer cells arise—the seed cells where genetic mutations that progress to cancer take shape. When a stem cell's DNA is overcome by a genetic coup d'état—brought on by hormones, radiation, toxins, viruses, and such—a mutinous army of immortal renegades arises capable of invading, metastasizing, and terrorizing everything in sight.

Invasion, *metastasis*, and *regenerative immortality* are the hallmarks of cancer—a *process* that proceeds in sequential, genetic steps from a normal cell to a malignant renegade. With their strong offensive line (invasion and metastasis), and their even more impressive defensive strategy (regenerative immortality), cancer cells carry with them the potential to outrun every race. Statistics bear out these gruesome truths: over the past fifty years, the War on Cancer has been long on treatments and short on cures. In most cases, the cancer victories are few and pyrrhic, and the truces, when they come, are often painfully temporary.

Chapter 2

Mendel

Gregor Johann Mendel was born and raised in the bucolic countryside of nineteenth-century Germany. There was nothing promising either in his childhood or in the early years he spent pursuing a vocation as a monk to indicate that he would change the very nature of biology. At a glance, Mendel appeared to be nothing more than a quiet, obedient cleric who liked to grow peas in the monastery garden, albeit quite a lot of peas. It wasn't until thirty years after his death that Mendel became a star. Today, words fall short of describing the importance of what this unassuming monk bestowed on us: the field of *genetics*.

Mendel was born in 1822 on a farm located north and east of Munich in an area that is now part of the Czech Republic. His family had lived and cultivated this same plot of paradise for more than a hundred years. The land there was similar to what we find in Lancaster, Pennsylvania, today: softly rolling hills, verdant pasture, excellent soil, hardwood and fruit trees, and a ready supply of water. As a child, Mendel kept a garden and tended bees—just what you might expect a farmer's boy to do. But as he reached maturity, Mendel believed that God had called him for other work, so he left his family and the farm and went off to become a friar.

In 1843, Mendel entered the Order of Saint Augustine Eremites, a branch of Augustinian clerics founded in 1256. Then, just as it is now, its mission was to encourage teaching and learning in art and science. Like all religious people, the Augustinians prayed at regular intervals throughout the day and kept to a strictly disciplined, prayerful life of poverty, chastity, and obedience. In addition, this order of Augustinians worked outside the monastery, typically as teachers in the local community.

Mendel entered Saint Thomas's Abbey at Brno to begin his training for the religious life. Brno was then a small village located at the confluence of two small rivers. Its name, translated, means "muddy, swampy, fortified hill." Mendel was ordained eight years later, in 1851, and sent to Vienna to study physics at the University of Austria. Vienna was an ancient, refined city located in the heart of Europe. As an epicenter for art and science, it must have been a marvel for young Mendel, who spent two years studying there before returning home to Saint Thomas's Abbey to teach physics at the monastery school. As might have been expected of a friar who'd come from a farming family, Mendel was allowed to plant a garden and raise peas, which, as history would later show, was like giving Michelangelo a ceiling, brushes, and some paint.

In addition to being a devoutly religious man, Mendel was a naturalist who was as interested in how God revealed himself through nature as how he revealed himself in scripture. Mendel was familiar with Charles Darwin's opus, *On the Origin of Species*, which was published in 1859, eight years after Mendel was ordained. Unlike the majority of theologians who decried the suggestion that God's creation was little more than a battle between circumstance and chance, Mendel was not put off by the theory of evolution. In fact, he thought the hypothesis both reasonable and sound. Mendel implied as much in the introduction to his first paper, written in 1865, concerning his experiments with peas. In that article, Mendel said that his goal was to put Darwin's theory to the test using plants in his garden and thereby demonstrate the "connection with the history of evolution in organic forms."[1] This level of open-mindedness was not exactly what one might have expected of an Augustinian monk, certainly not one like Mendel, who was so deeply embedded in the culture of nineteenth-century Germany.

In an attempt to understand the relationship between Darwin's theory and the divine will of God, Mendel used his garden as a kind of laboratory to work out the mathematical laws that governed evolution. This is exactly what Sir Isaac Newton did two centuries before when he used calculus to work out the laws of motion. Newton saw apples falling from trees and planets falling across the heavens, and he calculated the laws of nature that governed their motion. Mendel wanted to discover the natural laws that governed the theory of evolution—and express them mathematically—just as Newton had done with the laws of motion.

Mendel conducted hundreds of experiments with different kinds of peas in his garden, observed and recorded the results, analyzing them using the language of mathematics. He studied what the numbers had to say about what was going on. For the next eight years, Mendel crossed different types of peas in a process known as *hybridization* to see if he could discover the natural laws that would prove Darwin's hypothesis true.

Mendel began with the simplest experiment he could think of. He took a green pea and hybridized it with a yellow pea. Then he recorded the number and the color of the pea plants that were produced. Were they all green? Were they all yellow? Were there some of each? And if so, how many of each color were there? Mendel kept repeating the same experiment to see if he obtained the same results every time. He looked for patterns in the results he obtained. He analyzed the patterns. He asked himself, "Do these patterns fit into mathematical formulas that can be used to reliably predict the results of the next round of experiments?" If the answer was yes, then Mendel knew he'd hit upon a *natural law*, one that could be described mathematically, like Newton's laws of motion.

Mendel continued mixing and matching peas with different colors and different traits. Slowly but surely, the patterns of inheritance began to emerge from the data he collected, and these were used to generate formulas—mathematical models—that could be used to successfully predict the results of the next round of experiments. These mathematical formulas became known as the laws of inheritance, and Mendel believed that they supported Darwin's theory of evolution. Like Newton's laws of motion, Mendel's laws of inheritance accurately described and reliably predicted the heritable behavior of peas in his garden. But unlike Newton's laws, which applied to falling rocks, falling stars, the rising moon, and the setting sun, Mendel's laws applied to *living things*. No one had done *that* before!

Mendel didn't discover genes, per se—though he assumed that some quality or substance in the plants carried out the laws of inheritance. But his experiments implied that genes existed, for there had to be some means by which a trait (for example, the color of the pea) was carried from one generation to the next. The carrier of the trait was deemed to be the *gene*. Mendel didn't have to see the genes to know that they existed, any more than a driver needs to see the engine of a car to know it's there when she puts the key in the ignition, steps on the gas, and drives down the road. Nor does our hypothetical driver have to understand a single thing about the ideal gas law (PV=nRT) as it applies to the combustion engine to drive successfully and well. Mendel never saw a gene, but he saw what genes did, and he learned to predict what they would do next. This was more than enough.

With all this amazing data rolling in on peas, Mendel decided to expand his repertoire and try his hand with bees. He crossbred different strains, just as he had done with peas, in the hope that the results would be equally revealing. But Mendel was in for an unpleasant surprise. His bee experiments led to the emergence of a vicious strain whose queens were sexually insatiable—a bit much for a celibate, we can imagine. This cured Mendel of any further interest in playing around with bees. Stung by how quickly things got out of hand, he retreated to the safety of his garden where peas might grow but never move.

Eventually, Mendel obtained a set of mathematical laws that was as reliable in predicting the inheritance of traits as Newton's laws of motion were for predicting how long it would take for the next apple to hit the ground when falling from any height. When Mendel had finished his calculations, he discovered that each plant contributes half of the biologic information required for the physical expression of a trait—say, a green pod. This fact is the first law of inheritance. Interestingly, although each parent pea contributed half of the genetic information for a trait, some genes were dominant and some were recessive. If both parents contributed a "dominant" green gene, then all the plants in the next generation would be green. But if one parent contributed a "recessive" yellow gene and one parent contributed a "dominant" green gene, then the next generation would all be green, too. However, if plants from that hybridized generation were subsequently hybridized to yellow plants with only yellow genes, then that generation would have some green plants and some yellow plants—and this would occur *in mathematically predictable proportions.* Mendel discovered not only that each parent contributes exactly half of the genetic information for the expression of a physical trait but also that in a mixture of different genes, some are more dominant than others. And the dominant genes will always steal the show.

Encouraged by his early results, Mendel decided to increase the complexity of his experiments. He wanted to know if different traits—say, the color of the pea and the color of its pod—were inherited independently or if traits were linked together in some way. To sort this out, Mendel hybridized peas with two different characteristics (for example, green peas in yellow pods were hybridized to yellow peas in green pods). These more complex experiments revealed that traits are passed down independently of one another. This is exactly why people who are born with blond hair can have blue, brown, or green eyes and why people with blue eyes can have brown, red, or black hair or curly, straight, or thick hair. It's why you can have your mother's nose, your father's eyes, your grandmother's hair, your grandfather's jaw, and your Aunt Sadie's complexion. All traits come tumbling down through the generations completely independent of one another—to make the snowflake that is uniquely you. And that is the second law of inheritance that Mendel discovered in his garden.

Despite his disconcerting experiments with bees, Mendel continued to be curious about whether or not the laws of inheritance applied to animals. Next, he turned his attention to mice, which were everywhere and plentiful. Once again, Mendel's experiments were thwarted by the unforeseen. His bishop, Anton Ernst Schaffgotsch, learned that Mendel intended to breed mice and immediately halted the experiments. Apparently, Bishop Schaffgotsch was appalled at all the sex that would be required to generate sufficient data—or was he afraid of the results? After all, mice were mammals, just like people. We'll never know, except that Schaffgotsch went on the record as opposed to wanton sex, even if it was confined to rodents. Beaten back again, Mendel gave up all thought of experiments with animals and returned to the twenty-nine thousand pea plants growing in his garden.

In 1865, Mendel published his first paper. In presenting his laws of inheritance, he said he would leave the final judgment about his work to "the friendly decision of the reader."[2] Mendel's paper was well received, and the contribution he made to the field of biology was duly acknowledged among the small group of academicians who were familiar with his work. His scientific colleagues seemed to comprehend the importance of what he'd discovered, but they didn't give it much currency. Yes, it appeared that traits were shuffled and distributed in mathematically predictable patterns and then passed down through the generations. Indeed, these were the natural laws of inheritance. But, at the time, Mendel's scientific colleagues didn't know exactly what to do with these new laws, or how to hang them on the Tree of Knowledge without disturbing other, precious fruit. (Darwin's theory of evolution was shaking the boughs enough as it was.)

Two years after Mendel published his paper, he was made the abbot of Saint Thomas's Monastery and was forced to abandon his experiments for lack of time. He died of kidney disease in 1884. While the scientific community didn't really know what to make of the laws of inheritance during Mendel's lifetime, within twenty years, the theoretical and practical importance of his discovery began to dawn brightly on biologists around the world.

Chapter 3

Lathrop and Loeb

Without a hint of skepticism—at least nothing official can be found in the record of his presentation to the scientific community—those who were familiar with Mendel's work agreed with his conclusions: there were natural laws that could be described mathematically that governed the inheritance of traits in plants, similar to the laws of motion that governed the movement of planets across the sky. Scientists applauded the depth and breadth of Mendel's research (he had conducted hundreds of experiments), and then affably concurred that he had discovered the laws of inheritance, at least in peas. But as is often the case with a new discovery that does not drop neatly into a waiting niche within the body of knowledge, Mendel's papers were largely ignored.

Perhaps the world was too distracted by a darkening cloud looming over everything else, from biology to religion: Darwin's theory that species adapt to their environment and so evolve based on the survival of the fittest. Where was the divinely ordered, comforting hand of God in that? Where in this cosmic rodeo were the laws of Moses, the covenant of Abraham, the Golden Rule, or the rights of man if all of life, and our place in it, boiled down to something that looked suspiciously like sport?

Mendel's discovery that the inheritance of traits in plants was reliably expressed as a set of mathematical probabilities may have been perceived as yet more discomfiting news. Indeed, it must have felt like the old, tight picture puzzle of the world, which had taken thousands of years and thousands of stories to build, was falling apart, taking divine order with it, while the new pieces that Darwin and Mendel brought to the table didn't fit anywhere at all.

Meanwhile, in the not-too-distant background was the unspoken threat that if Mendel's laws were put to the test and proved true in animals, then humans would not likely be found exempt. No doubt, many scientists, who were as religious as anyone else, must have felt that the bindings were coming off their textbooks, papers flying everywhere. If Darwin and Mendel were correct, then God had created a world that was nothing more than a wild, uncertain, unfair arena where he was little more than a spectator to the circuses of beasts, including people. These unexpected revelations must have sent jolts of doubt blistering through even the most intrepid minds. How could these discoveries, verified by scientific experiment, be reconciled with faith in the divine providence of God? Thus, after centuries of ordering everything neatly into place, science and religion seemed to be coming apart at the seams.

At the beginning of the twentieth century, scientists looked with a fresh eye at Mendel's papers. Immediately, they recognized in the laws of inheritance the opportunity to identify the genes that cause cancer. They would start their search by looking at families and communities where the disease was found to be most common (in other words, where the disease appeared in clusters). But little did they know that in their quest to find the genetic cause of cancer, they would charm from the basket not a handful of cancer genes but a cobra—the breast cancer virus.

***

To continue the story, we have to return for a moment to the middle of the nineteenth century, in rural Illinois, 1868—the year Abbie Lathrop was born. Lathrop's parents were both teachers who homeschooled Abbie, their only child, until she was sixteen. Abbie left home for two years to get a teaching certificate and returned to Illinois to take a position at an elementary school close to where her parents lived. But Lathrop, who suffered from pernicious anemia, didn't have the stamina to teach.

Pernicious anemia is a chronically debilitating disease that begins in a group of cells that line the stomach called *parietal cells*. Under normal conditions, parietal cells absorb vitamin B12 from the food we eat. Patients with pernicious anemia have nonfunctioning parietal cells that can't absorb vitamin B12. Because every cell in the body requires vitamin B12, a deficiency of this essential vitamin results in a host of chronic symptoms that slowly destroy the health of the patient. Eventually, the patient with pernicious anemia becomes totally incapacitated and a protracted death ensues, usually in midlife.

Like in most patients with pernicious anemia, Lathrop's disease produced one strange malady after another, each one rendering her increasingly debilitated. As a result of the unrelenting and progressive nature of her disease, Lathrop was forced to retire from teaching after only a few years. However, she did not lose her enthusiasm for keeping busy. In 1900, at the age of thirty-two, with her mother and father both deceased, Lathrop left Illinois and moved to her parents' hometown of Granby, Massachusetts. She bought a chicken farm, a business that she must have thought would be easy enough to manage since she could hire laborers to do the heavy work of feeding and tending the flock.

But the chicken farm didn't work out either. Lathrop eventually sold the birds and switched to raising small animals to sell as pets, employing local children to help with the chores after school. It was a small operation, but it provided enough income to pay the bills and keep things going. Though Granby was little more than a village at the time, the big city of Boston wasn't far away. In fact, Boston was near enough to be a ready market for Lathrop's little pets: guinea pigs, rabbits, ferrets, canaries, and *mice*.

Yes, mice. Hard to imagine, but having mice as pets was an increasingly popular hobby at the turn of the century. It originated in seventeenth-century Japan, where mice were bred for appearance and behavior, just as we breed dogs today. Coat color (cream, white, lilac, yellow, or sable) and eye color (pink or ruby red) were seen as the most desirable features. Especially prized were "waltzing" mice, a strain that appeared to dance around because of an inherited defect of the inner ear that affected their sense of balance. The craze for keeping mice as pets spread from Japan to China, and then, with the shipping trade, from China to England and on to America. Pet mice and all the trappings—clubs, newsletters, breeders, and conventions—were popular in the United States during the late nineteenth and early twentieth centuries, especially in New England.

Lathrop began her mouse-breeding enterprise by purchasing a pair of waltzers; nature took care of the rest. Her new business, though modest, brought a steady income and was increasingly successful. Finally, Lathrop was able to pay her bills, turn a comfortable profit, and keep the local children busy earning tiny, happy salaries. Lathrop soon expanded her product line, advertising in local newspapers so she could acquire other varieties of mice to expand her inventory. Soon mice fanciers, as they were called, began to seek out her expertise. In no time, Lathrop's business turned into a thriving cottage industry. Then a few years later, her little farm became a bonanza when scientists from Harvard University in Boston, driven by the quest to use Mendel's laws of inheritance to find the genetic cause of cancer, drove out to her farm to buy her mice in bulk.

To meet the demand from the scientific community, not only in Boston but also around the country, Lathrop expanded her inventory to eleven thousand mice. It must have been a sight! Researchers converged on her farm and drove away with trunkloads of rodents for their laboratories. Lathrop enlarged her work force, paying her little helpers seven cents an hour to feed the mice and clean their cages. All the while, Lathrop's reputation grew. And so did her income. In 1900, a typical salary for an

American worker was $450 a year. A teacher would be paid, on average, $1,000 a year. Lathrop was making twice that: $2,200 a year—decidedly more than many of her academic clients.

Lathrop was not only industrious, she was also alert and curious. She wanted to know what, exactly, the scientists were up to with all these mice. The men from Harvard University informed her that they were conducting experiments aimed at discovering cancer genes as a way to find the cause of cancer. Lathrop mentioned to one of them, Clarence Cook Little, that if he was interested in finding cancer genes, he might want to take a look at *her mice with breast tumors*. Little must have lit up like a Christmas tree when Lathrop showed him her mice with breast tumors. Taking the gleam in his eye for a jewel in her hand, Lathrop quietly lit up, too. After sending Little on his way with some of her mice with breast tumors, she promptly wrote a letter to one of her other clients, Leo Loeb, MD, of the University of Pennsylvania, offering to send him some of her tumor-prone mice so that he might "determine the cause of the malady." You see, at that point, Lathrop only *suspected* that her mice had breast cancer. She needed to know for sure.

Loeb was a pathologist, and Lathrop knew that he had a special interest in cancer. Loeb replied immediately: yes, he would be delighted to perform autopsies on her mice to determine if they had breast cancer. Lathrop shipped her tumor-prone mice to Philadelphia, and Loeb returned the diagnosis: *Lathrop's mice had breast cancer*. Loeb then suggested that he and Lathrop collaborate, for there was work to do! Loeb offered to design experiments for Lathrop to carry out on her farm, using a corner of her shed as a laboratory. In a stroke of equality well ahead of his time, Loeb proposed that they collaborate as equals and share joint authorship on all papers published as a result of their efforts. Having never met Lathrop, who had no academic credentials to speak of, Loeb's offer was quite the leap of faith. We'll never know what, if any, doubts Lathrop harbored about this new undertaking for which she had no training, but she readily agreed to the arrangement and waited for further instructions. With the exchange of these two letters, Lathrop underwent a metamorphosis that transformed her from an undistinguished spinster with a pet farm to a capable scientific investigator and skilled mouse surgeon. This unlikely collaboration between very unequal strangers proved to be a historically important moment for breast cancer research. Even more amazing was that it was set in motion by a solitary, industrious woman who used little more than ink and pen and paper to blaze a trail that we're still following today.

The experiments that Loeb designed in Philadelphia and that Lathrop executed on her farm in Granby were meant to answer a series of simple questions, each considered one at a time, that, hopefully, would illustrate the natural history of breast cancer in mice. Lathrop carried out Loeb's carefully designed experiments, many of which involved intricate surgery on tiny mice, recorded the data, and then sent the results to Loeb for analysis. He would then compile the results, write a scientific paper, include Lathrop as a coauthor, and submit their article to a peer-reviewed journal.

As is the case for all research, the results of one experiment were used to inform the next, with each additional experiment designed to increase the body of knowledge around one particular subject, one answer at a time. Loeb and Lathrop, two of the most unlikely partners who ever worked together—and never met—conducted a series of fascinating experiments on mice with breast cancer over the next eleven years and published their results in medical journals between 1907 and 1918. These articles broke new ground on our understanding of breast cancer, and the discoveries that issued from Loeb and Lathrop's experiments remain as profoundly significant and applicable today as they were a hundred years ago.

What, then, did these unlikely investigators discover about breast cancer in Lathrop's mice? First, they documented what Lathrop had observed when she first started buying and breeding mice: some strains of mice had higher rates of breast cancer than others. In particular, a strain of mice with "lethal yellow" fur was found to have a high incidence of breast cancer, while another strain with black fur seemed to be free of the disease entirely. Other strains also developed breast cancer—but only infrequently, and only when the mice were much older. Loeb and Lathrop concluded that because some strains of mice had high rates of breast cancer, some had low rates, and some appeared to be entirely immune to the disease, genes must play a role in causing breast cancer in certain mice.

Clearly, this was an important place to start: there are breast cancer genes in mice with breast cancer. In fact, it is the bedrock of our understanding of breast cancer in mice and women today. And because they are genes, albeit cancer genes, they ought to obey the laws of inheritance set forth by Mendel, just like the genes for eye color do.

In an article published in 1915, "Further Investigations on the Origin of Tumors in Mice," Loeb and Lathrop reported that in strains of mice with a *higher incidence* of breast cancer (for example, the lethal yellow), their tumors tended to develop at *an earlier age*. Interestingly, these mice also tended to enter puberty at an early age, too. In strains of mice with a *low incidence* of breast cancer, their tumors developed when the mice were *older*. They also started *puberty at a later age*. Loeb and Lathrop concluded that the risk for breast cancer, and the age at which it occurred, represented "distinct factors which frequently, but not in all cases, are in some way (related) to each other."[1]

But how were risk and age and puberty related? Did the low-risk mice carry fewer breast cancer genes than the high-risk mice? Was this purely a question of quantity—in the same way that the color of icing is a question of the amount of food coloring that is added to the mix, with a single drop of Red No. 3 producing pink and three drops yielding an eye-popping red? Or did both the high-risk and low-risk strains carry the same number of cancer genes, with puberty acting like the accelerator in a car? If puberty occurred later, the accelerator pedal was merely tapped, and the tumors appeared in smaller numbers and at a later age. But if puberty occurred earlier, the accelerator pedal was pressed to the floor, and tumors occurred more frequently and at an earlier age.

Or did both these factors, cancer genes and puberty, embrace and court and dance around each other over time (age, that is) in ways too complicated to be discerned from these preliminary, simple experiments?

Loeb and Lathrop weren't sure, but the results clearly demonstrated that cancer genes and the time of puberty and the onset of cancer in cancer-prone mice were linked, and locked, and interactive.

Loeb then wondered, if puberty played a role in stimulating the genes for breast cancer, did pregnancy play a role, too? With this question in mind, Lathrop conducted another experiment in which the influence of pregnancy on the rate of breast cancer in high-risk mice—the ones presumed to have a greater number (in other words, additional "drops") of cancer genes—was examined. In this experiment, pregnancy did, indeed, increase the rate of breast cancer. Loeb and Lathrop reported that breast tumors occurred more frequently in female mice that were allowed to breed (and increased even further as the number of pregnancies went up) compared to female mice that were kept aside as "vestal virgins." Ah, so puberty and pregnancy both played a role in increasing the risk for breast cancer in breast cancer–prone mice—the ones with the cancer genes.

This experiment strengthened the argument that the female reproductive cycle played a powerfully influential role in controlling the action of breast cancer genes. Indeed, in male mice there was rarely a breast tumor to be found, even though males from the high-risk breast cancer strains clearly carried the same cancer genes as their female relatives. But wait: male mice didn't get breast cancer even though they carried the same breast cancer genes—but they must, because that is part of the laws of inheritance. What? Why not? Certainly, if Mendel's laws applied to breast cancer genes, then males carried their fair share (in other words, an equal portion). But Lathrop reported that males from the high-risk strains did not develop breast cancer: it was only the females with their functioning ovaries (puberty) and their pregnancies that experienced the full force of the breast cancer genes. Males certainly carried breast cancer genes in equal measure, but it was the females who lived with the consequences—consequences amplified by puberty and breeding.

Loeb and Lathrop then discovered that mice could be prevented from getting breast cancer if their ovaries were removed before puberty. Wow! No ovaries, no breast cancer—not even in high-risk mice. Scientists had known for years that in humans there was a "mysterious sympathy" between the reproductive organs and the breasts, but Loeb and Lathrop were among the first to demonstrate a direct relationship between the ovaries and the risk for breast cancer. Loeb and Lathrop had no idea that hormones—particularly estrogen—were the source of the fuel that feeds breast cancer genes, for no one at the time was aware that hormones even existed—though many scientists suspected that the ovaries produced "secretions" that directly impacted breast tissue. It would be another thirty years before Adolf Butenandt, a Nazi scientist, discovered that the ovaries produce and secrete estrogen, a hormone that, at puberty and thereafter, stimulates the growth of breast tissue and influences all of the activities that take place there—including activities like cancer.

Loeb and Lathrop published their last paper together in 1918, the year Lathrop most likely died of pernicious anemia. (She was fifty-one years old at the time.) For their final experiment, Loeb and Lathrop crossbred high- and low-risk mice to see what would happen to the rates of breast cancer when high-risk males were mated to low-risk females, and vice versa, when low-risk males were mated to high-risk females. They reasoned that if cancer genes behaved like genes for, say, eye color, then they would obey Mendel's laws of inheritance; and the results obtained from crossbreeding high- and low-risk mice would be as predictable as those obtained from hybridizing peas. For example, if a high-risk male was mated to a low-risk female, the daughters of this match should have a rate of breast cancer that would be a *predictable, reproducible balanced mix between the two.* If cancer genes behaved like other genes, then both the male and female mice would carry these genes in such a way that they would *be passed on proportionally to the next generation.*

But remember, Loeb and Lathrop had also discovered that puberty (in other words, the ovaries) and pregnancy played a role in influencing the risk for breast cancer. And these were strictly female factors. So, were the breast cancer genes strictly female, too? They weren't sure. Because of the questions raised by their discovery that female factors altered the risk for breast cancer in high-risk mice, Loeb and Lathrop decided to go back and take another look at the inheritance patterns of cancer genes in high- and low-risk mice strains to prove that these cancer genes did, in fact, obey Mendel's laws of inheritance.

What did Loeb and Lathrop find when they crossbred mice from high- and low-risk strains? Did they find what Mendel had found with green and yellow peas? No! They found that it was *the females from the high-risk strain* that passed the risk for breast cancer on to their female offspring. The males from the high-risk strain passed on a little risk, but not that much. Additionally, they discovered that the males from the high-risk strain imparted a slightly greater risk of breast cancer to their female offspring if they were mated to low-risk females but nothing compared to the increased risk for breast cancer observed when high-risk females were mated to low-risk males. Again, there was something about those females, something about the action of breast cancer genes in female mice from high-risk strains. When a low-risk male was mated to a high-risk female, the incidence of breast cancer took off like a rocket—far in excess of that predicted by the laws of inheritance. The female mice from the high-risk strain transmitted the risk for breast cancer in a profound and durable way—even allowing for the "accelerator" action coming from the ovaries and pregnancy. There was only one conclusion to be drawn: breast cancer in Lathrop's mice *absolutely did not* obey Mendel's laws of inheritance—and the females from the high-risk strain were the ones to blame!

Lathrop found that the female mice in the high-risk strains did not contribute half the risk for breast cancer—as Mendel would have predicted—but almost all of the risk! The experiments indicated, time and again, that the risk for breast cancer was passed strongly down the maternal line, while the risk attributable to males was present but relatively lame. Either Mendel was wrong about males and females carrying exactly half the genetic load for every trait or Mendel was correct and *there was something else* afoot. To make things even more confusing, Lathrop found that once the risk for breast cancer had been transmitted to offspring as a result of crossbreeding a *high-risk female* with a *low-risk male,* the increased risk for breast cancer was preserved in every subsequent generation. Once it was "in the door," it stayed. There was no way to get rid of the increased risk for breast cancer once it had been introduced into a "family tree." Even more bewildering, Lathrop found that once a new high-risk strain was produced by mating low-risk males to high-risk females, further mating with low-risk males did not dilute the risk for breast cancer whatsoever. Once the risk for breast cancer was introduced into the female line, the risk for breast cancer could not be diminished by subsequent mating with low-risk males.

If you find it difficult to follow the thread of the preceding paragraph, you can find comfort in this: Loeb and Lathrop, and many other scientists who followed in their footsteps over the next thirty years, were equally perplexed. In fact, they were stumped. And they remained so until one brilliant scientist, John Bittner, explained the crazy tango that was taking place between the inherited breast cancer genes, the female factors (hormones and pregnancy), and a little something else he found—something swimming around in the milk of high-risk females.

Using only the most rudimentary materials at hand, Loeb and Lathrop opened up a whole new frontier for breast cancer research. Furthermore, they were well ahead of any progress being made in breast cancer by the academic titans not far away at Harvard University. Loeb and Lathrop documented the existence of breast cancer genes—yes, they did exist; and they identified the strains of mice that carried them and those that did not. They showed that the ovaries and pregnancy play an important role in influencing the action of these breast cancer genes, and that removing the ovaries and preventing pregnancy either entirely prevented or significantly reduced the incidence of breast cancer in high-risk mice. They discovered, to their surprise, that the high-risk females were the ones primarily responsible for passing the risk for breast cancer to their offspring, and that once the high-risk females introduced breast cancer into a family line, it became the dominant feature of every generation that followed. And although the results of their last experiment produced a tight, wet knot that refused to come untied for the next thirty years, what Loeb and Lathrop discovered about the natural history of breast cancer is as applicable today—in mice *and women*—as it was a hundred years ago.

Chapter 4

Little and Jackson

Clarence Cook Little—everyone called him Prexy—was one of Lathrop's most reliable customers. Little was a fifth-generation direct descendent of Paul Revere who enrolled at Harvard University in 1906 to study biology. You might say that this patrician son of an old Boston family was a born geneticist. His family had been breeding cocker spaniels, Scottish terriers, and dachshunds for years. He'd been given a pair of pigeons at the age of three and had won first place for another pair he bred at the age of seven. Little was introduced to William Castle, a professor of biology, during his first year as a Harvard undergraduate. Castle was an enthusiastic supporter of Mendel's work and was determined to test the laws of inheritance in animals—the precursor to understanding how they worked in man. Little was very familiar with the practical application of Mendel's work as a result of his experience breeding pigeons and dogs, even if he wasn't entirely clear about the mathematical application of the laws of inheritance to mammalian genetics. His mind was primed for learning more, and Castle was more than happy to take him under his wing.

Little was eager to help Castle pioneer the uncharted territory of mammalian genetics. Thinking he was the man with superior knowledge about how best to proceed, Little suggested to Castle that they breed dogs for the study of the patterns of inheritance in mammals. But Castle was the smarter man. He pointed out, quite rightly, that dogs were large, expensive to feed and breed, took years to come to maturity, and produced only one litter every year, if that. Castle pointed Little in the direction of mice: they were smaller, cheaper, reached maturity earlier, produced three litters (on average) every year, and they were mammals, very much like man. Castle then offered his strongest argument in favor of mice: more experiments could be run in a shorter period of time, which would translate into the production of more scientific papers, which would increase their competitive edge in the growing field of mammalian genetics sprouting up on every major campus around the country. By breeding and studying mice instead of dogs, Castle and Little could establish themselves as leaders in what was sure to become the next new thing in biology. Little immediately understood the wisdom of Castle's suggestion. He abandoned all thought of dogs—except that he continued to breed them as a hobby for decades more—became a convert to the use of mice, and headed out to Lathrop's farm to get some. Little came back from Granby with boxes of rodents, bred them in a small office at Harvard University, and by 1909 was known all over campus as Castle's "mouse man."

The initial batch of mice from Lathrop's farm reproduced at a rate commensurate with their reputation—fast and furiously. As a result, the small space that Castle had provided in the Department of Biology was soon teeming with the Little's varmints. The move to Harvard's Bussey Institute for Agriculture, located in the countryside outside Boston not far from Lathrop's farm, was a welcome relief for all concerned. While Little was busy breeding mice to ferret out the secrets of mammalian genetics, he came up with a brilliant idea worthy of his revolutionary bloodline. The idea was to inbreed mice over many generations in order to create strains—that is, purebred strains—that would be genetically identical. Inbred mice are created by breeding siblings (brothers and sisters) repetitively. By concentrating the gene pool over many generations, eventually all the mice would share identical genes. Strains of genetically identical mice that, say, all shared a particular trait worth studying, such as obesity or cancer, would allow Little to isolate the genes responsible for producing that trait. Once the trait was made common in an inbred, purebred strain of mice, and the gene responsible for the trait identified, the experiments could be conducted that would be able to test one treatment or another to see what, if anything, altered the onset or progression of disease.

Little was an ambitious man with a leonine goal. More than anything, he wanted to isolate the genes that caused cancer. If he was able to create strains of mice that were *genetically identical for a specific cancer trait*, he would be able to conduct experiments on his "cancer thoroughbreds" confident that the results would derive entirely from the intervention that was being tested (for example, a drug, a carcinogen, or a surgical treatment) rather than as a consequence of other—random, unseen, genetic—factors. Little envisioned the creation of genetically identical, purebred mice that would be the key to isolating cancer genes and finding ways to stop them in their tracks. If Little achieved his goal, he would transform experimental mice from practical to precious. He would revolutionize cancer research and set his mark upon the world.

Little's idea was brilliant, but it was also flawed. Everyone knew, including Little and everyone in his family, that inbreeding weakens the strain. Horse breeders and dog breeders knew only too well that breeding brothers and sisters in an effort to highlight one desirable trait inevitably gives rise to unexpected, debilitating flaws that mar the strain, often fatally. Look what inbreeding, of a sort, had done to the thrones of Europe. The marriages of royal cousins had filled these seats of power with hemophiliacs. With horses and dogs, it had been seen time and again that inbreeding resulted in strains in which few animals survived to maturity; and those that did often were infertile. Nature adores diversity. Evolution demands it. A consolidated gene pool is a stagnant, lifeless place. No doubt, Little knew he was in for trials in trying to render a viable strain of inbred mice.

And one of his first trials was overcoming Castle's understandable skepticism. Not only did he think Little was doomed to fail, he thought it was a waste of time to try.

Undaunted by the formidable obstacles that stood in his way—his mentor's stern rebuke, the forces of nature that killed off anything that tried to overrun the system—Little set out to prove his case. He would succeed where others had failed. He would be "the man." Little fell in love with his idea for another important reason: the creation of genetically identical inbred mice would ensure the replication of experimental results across the board. Using genetically identical mice would allow scientists at one institution to reproduce the experiments reported at other institutions, with a guarantee that all results would be the same. Little understood that if he could standardize the animals used in experiments, then experimental results would be reproducible in any laboratory, and, therefore, the conclusions drawn from those experiments would be far more reliable. He had nothing to lose—he was still only an undergraduate, after all—and everything to gain; so he pressed ahead with the exuberant confidence of youth and inexperience.

Luck was on his side. That and more luck. Within two years, Little had produced a strain of viable, vigorous, fertile inbred mice. Having won the day, his horizon opened circumferentially. And so did Castle's for having such a bright young man in his department. Even if Castle harbored slight chagrin that his whippersnapper student had forged ahead and proven him wrong, the upshot was that the old professor was now bathed in a sparkling light. Viable, genetically identical strains of mice streaming out of the Bussey Institute on Castle's watch were Little's achievement and Castle's glory.

Little remained at Harvard University for the remainder of his education, receiving a baccalaureate degree in 1910, a master's degree in 1912, and a doctorate in 1914. Before he was halfway through, he and his thoroughbred mice had become scientific celebrities. His barnstorming trick with inbred mice quickly led to the expectation that the cures for all kinds of diseases would be found in the study of these mice. As Little had made clear early on, though, the disease that he wanted to nail most of all was cancer.

Beginning with a handful of Lathrop's mice—the ones with breast cancer, that is—Little created an inbred strain (C3H) in which over 90 percent of the animals developed breast cancer before they died. The C3H strain proved to be so stable and prolific that it soon became the favorite animal for breast cancer research and remains so today. Other institutions quickly followed suit. Everyone began trying their hand at creating inbred mice. But Little had a head start and he intended to keep it.

War broke out in Europe in 1914, the year Little finished up at Harvard. He enlisted in the air force, hoping to become a flying ace, but he was assigned a desk job instead. After completing his tour of duty, Little spent some time working at the Carnegie Institute's Station for Experimental Evolution in Cold Spring Harbor, New York, where, it was reported, "he spied on rats and mice." Finding the genetic cause of cancer, and then its cure, were the foremost tasks on his mind. But Little did not intend to do this singlehandedly, in a laboratory, slaving away over one experiment after another. He intended to become an emperor of mammalian genetics, leading an army of cancer investigators at some major institution. The Carnegie Institute in Cold Spring Harbor was a first step along the way; it was not, by any means, a destination.

In 1918, while Little was still in uniform, Lathrop died of unknown causes that were likely related to the accumulating effects of pernicious anemia. But she, too, had served her country well. During World War I, the US military purchased Lathrop's mice and employed them like canaries in a coal mine to detect the presence of mustard gas on the battlefields of France.

In 1906, when Little entered Harvard, cancer was still as mysterious and terrifying as the Black Death had been five hundred years before. It was a disease that seared the souls of victims and onlookers alike. Nothing—not even the centuries of renaissance, reason, and enlightenment—put a dent in the miseries associated with malignancy. Ancient doctors had tried applying caustic poultices. Some intrepid surgeons tried cutting the tumors out. Nothing worked. In virtually all cases, patients died more slowly and horribly than any torturer could imagine. At least something as dreadful as the bubonic plague was merciful: its victims were thrown into a stupor within days of getting sick and died in a state of oblivious prostration. Not so with cancer. It often grew and spread for years before dealing a fatal blow. Little realized that whoever discovered the genetic cause of cancer would live out his days in glory and would likely be in a good position to drill down on a cure—or so he hoped. Having successfully completed the first step on his way to a legacy by successfully creating inbred strains of breast cancer–prone mice, Little had every intention of riding his thoroughbreds to find those cancer genes.

While he was working at Cold Spring Harbor in New York, Little met Margaret Sanger, the woman who pioneered a woman's right to birth control. He immediately became one of her strongest advocates. In 1921, Little helped Sanger create the American Birth Control League. It was a bold and visionary move, for it flew in the face of conventional wisdom about the absolute rule of husbands in their homes. Little could have remained cordially supportive of her cause, but he was most outspoken. The "indiscriminate growth of families" got in the way of progress for men and women alike. It was a brave and revolutionary cry at the time, one in keeping with the bloodline he could trace back to Paul Revere.

The few years that Little spent at Cold Spring Harbor Laboratory served his curriculum vitae well, but he had no desire to remain at the bench as a "lab rat," sorting through mice for the rest of his career. In 1922, just eight years after receiving his PhD, Little became president of the University of Maine. Professionally, it was a giant step. But it landed him in a small place that he did not like, and which did not like him. The University of Maine was located in Orono, population 1,100. Coming from New York (population 3.4 million) by way of Boston (population 560,000), Orono must have felt culturally barren. With winter lasting a good eight months of the year—and spring a day or two—the environment honed a talent for endurance and not in an open-minded way. Little liked to say that experiment and change were the heart of life, but few in Orono agreed. Little found scant support for higher education in Maine, with the consequence that its residents were rather parsimonious about funding the state university.

Running a university was a job for which Little was neither prepared nor suited. But he didn't find that out until he left the University of Maine for the University of Michigan in 1925, thinking that all he needed was a change of scenery and a more liberal canvas upon which to make his mark. He didn't last any longer in Ann Arbor than he had in Orono. There were men on the board of regents and in the state legislature in Lansing who found his views on birth control deeply disconcerting and unbecoming of a university president. They made their opposition painfully obvious when Little submitted a budget of $4.9 million in 1927 to run the university and received a check for $1.3 million instead. A very wealthy alumnus, Fred Warren Green, donated $1.7 million to the university (more than its budget for the year), fully expecting that he'd be able to say exactly where the money was spent. Little didn't think so. In no time, everyone locked horns. He even got into tussles with the students, calling in federal agents to monitor the drinking on campus. He wanted all undergraduates to reside in the university's dormitories, infuriating thousands of landladies whose livelihoods depended on renting rooms to students. Little had had enough, and in no time the University of Michigan had had enough, too. He was not a born leader and he appeared to be chronically inattentive to the acquisition of skills that might make him one.

In January 1929, Little admitted defeat. He resigned as president of the university, saying that he hoped to "be more effective in scientific research and teaching than in administration."[1] Little offered to leave his post in September but was invited to depart in June. Sadly, his marriage had fallen apart, too. The details of the disintegration of his home life are murky and unseemly, with extramarital women implicated in the drama. In the end, Little left behind all he had, including three children and $100,000, and headed out of town with one of the female understudies in tow. The world was in pieces at his feet, but stepping over the entire mess, he moved on. He had made friends with many important men (and a few important women) over the years. His exit strategy was years in the making, with one man in particular acting as his most promising savior from a life of administrative drudgery for which he was not fit nor suited: Roscoe Bradbury Jackson, founder and president of the Hudson Motor Company.

Jackson was an engineer from the University of Michigan who'd made his fortune manufacturing cars in Detroit. He was one of the many wealthy tycoons who kept summer homes in Maine. During his time in Orono, Little made sure to introduce himself and pay his respects when these important men came east to escape the Midwest's summer broiling pan and find relief in the icy breezes sweeping inland off the North Atlantic Ocean. (The result was rather like removing your head from a blazing oven and sticking it in the freezer—a delightful change, at first, until you noticed how very cold it was.) Tycoons of the caliber of Andrew Carnegie and John D. Rockefeller set the standard for what should be done when a man became insanely rich: give much of it away. These were Jackson's role models, and they set the tone and pace for philanthropy. Any man who aspired to great wealth had to also aspire to great largesse. Jackson was looking for interesting ways to seed his money and was particularly taken with the success he had enjoyed using modern methods of industrial production. Like a boy with a drumstick, he looked for other reverberating surfaces on which he could beat out his melody.

Jackson kept an eye out for opportunities he thought worthy of his support, especially ones that seemed to fit his point of view. Over the years,

Little had succeeded in persuading Jackson that they could make music together with mice. Little's vision of using inbred strains of mice to find the genetic cause of cancer appealed to Jackson's belief that industrial engineering was the key to success on a grand scale. Grandness was one thing they both had in common. That and vision. Jackson's success with assembly line production of cars in Detroit had turned his conviction into a doctrine. He was eager to apply the principles of industrial production to philanthropic enterprises like curing cancer. Little readily agreed to the wisdom of employing Jackson's business models to making mice. By enthusiastically acknowledging Jackson's suggestion—nods embellished with a touch of flattery—Little acquired a valuable tip and gained a valuable, moneyed friend. They kept in touch. Indeed, when Little decided that he was ready to get out of Maine, he allowed Jackson to pave the way to the presidency of the University of Michigan, Jackson's alma mater. This move came with a thrilling, skillfully negotiated bonus: a president's research laboratory was created just for Little, a place where he could formally resume his investigations of mammalian genetics using his inbred mice. In addition to his handsome salary, Little was given enough money to hire seven graduate students to staff his laboratory—a young and eager strike force hungry for deployment on any mission that might strike Little's fancy. One of the brightest and most capable of these few good men was John Joseph Bittner, a geneticist and cancer biologist who was very interested in tumors, especially in mice.

Little was happy to have his president's research laboratory, but was less than happy with how little time he had to oversee its work. More and more, he realized that what he wanted was to run his own, large, prestigious research lab. That, however, would take a fortune—the kind of money that was within his sight and not his grasp. When things started to tank (again) in Michigan, Little began acting on his dream to head up his own cancer research laboratory. He reached out to Jackson, a man who had already donated funds to Little's research laboratory at the University of Michigan.

One can assume that Jackson was pleased with what Little had done with the money, pleased with the results of his research. Jackson agreed to help Little find the money he would need to build a first-rate research laboratory, one where the principles of industrial engineering would be used to discover the genetic cause of cancer—prelude to a cure. But Jackson wasn't going to pick up the tab entirely on his own. Either he wasn't in a position to write a check that would cover the costs of the entire project, or he didn't care to foot the bill without spreading some risk around. Instead Jackson offered to gather a group of philanthropic investors to fund the enterprise. This lightened the burden of Jackson's financial commitment to the project and gave his friends the opportunity of participating in the discovery of scientific breakthroughs, which, if successful, would lead to a cure for cancer.

In 1929, the year Little was canned in Michigan, Jackson had created a coalition of donors committed to building Little's research laboratory. Everyone decided it should be located in Bar Harbor, Maine. Back to Maine? Yes, back to Maine. It was the perfect location. The wealthy Detroit donors who'd backed the project had summer homes in Bar Harbor. It was a place where everyone would be at their leisure for weeks on end, with plenty of time to oversee, reflect, and contribute their two cents to the work Little was doing there. Maine may not have been Little's first choice, but Bar Harbor was an aristocratic cut above Orono, and it was an easy drive down to Boston and New York. Jackson and his business friends also reckoned that building a laboratory in Maine would cost less and free up more of their money for the research itself.

So 1929 started out badly for Little and then took a hard right and promising turn that spring. He would be leaving academia for good to embark upon his life's dream—heading up his own cancer research laboratory. Powerful, influential, and wealthy men had gotten behind him to make a way before him. Happy days were his again. That same year, Little became the managing director of the American Society for the Control of Cancer, the organization started by surgeons in the late 1800s to teach doctors about how to diagnose and treat cancer—not that there was much anyone could do at the time. When Little took the helm, the American Society for the Control of Cancer was the largest of its type but still quite small in terms of its impact on the community. Their insignificance would change when, in the 1940s and under new leadership, it became the American Cancer Society.

Over the years, Little had done an excellent job angling the media to highlight his use of inbred mice to find the cause and cure for cancer. Having turned their attention in the direction of Little's banging drum, the media followed his career closely to see what, if any, progress he was making with his thoroughbreds. The sudden, unexpected news that the University of Michigan had tossed him out of his presidency made sensational headlines and forced Little to find a clever spin for his woes. He was a cat with nine lives that knew how to spin around in midair and land on his feet. One month after submitting his resignation (February 1929), Little managed to make the cover of *Time* magazine looking very much like Clark Gable, but without the smirk. The story, "Jobless Little," carried the cover headline: "More games, fewer babies." The article discussed his involvement with the American Birth Control League, describing him as "a man of science, keen, enthusiastic . . . energetic and determined."[2] A detailed chronology of the various scraps he got himself into at the university concluded with, "Dr. Little, like most convinced men, worked fast, sometimes ruthlessly."[3] The implication was that he meant well. The *Time* article noted that once Little was safely on his way out of Ann Arbor, the board of regents spoke generously of "'his high ideals of educational standards,' 'initiative,' 'constructive aspirations,' 'frankness,' 'courage,' and 'sincerity.'"[4] In May 1929, as speculation about where Little would go next mounted—rumors circulated that he was to become president of Harvard University or that he might join the Rockefeller Institute—the *New York Times* broke the news that he had embarked on a joint venture with philanthropists from Detroit to create a biologic cancer research laboratory. In the *Times* article, Little made it very clear that he was intent on finding the *cause of cancer*. The unspoken promise was that once he'd discovered the cause of cancer, cures would be found.

At the time, there was little hyperbole in making an assertion such as this. Scientists had good reason to believe that finding the cause of a disease, say, smallpox, was one small step away from curing it. Nail the cause—a virus—and develop a vaccine; and so, it's cured. Based on recent experience and successes, the public had reason to believe that if the cause of cancer was found by, say, isolating cancer genes, then finding a cure would be a cinch. (Unfortunately, they were mixing apples and oranges. It is possible to make a vaccine against a virus, even a tumor virus, and so to prevent the disease. To date, it has been impossible to make a vaccine against a cancer gene—a trickier feat, by far.)

The cover story in *Time* smoothed out the more public wrinkles, and then things improved considerably in the spring when Little closed the deal with Jackson and reported the feat to the *New York Times.* But tragically, just a few months before the research laboratory was set to open in Bar Harbor, Jackson fell ill with influenza while vacationing in France and died. His sudden death at the age of thirty-four was a shock that rattled everyone. The other donors unanimously agreed to honor Jackson's memory by naming the center the Roscoe B. Jackson Memorial Laboratory. *JAX* was its Western Union address, and in an endearing slang that has come down through the years, it became known affectionately as JAX Lab, and its mice—descendants of those from Lathrop's farm—became known as JAX mice.

Jackson's death in March 1929 was a blow, but it was not severe enough to stop the show. That occurred six months later when the stock market crashed on Wall Street. The funds pledged to the Roscoe B. Jackson Memorial Laboratory simply no longer existed. The banks were bust. The money was gone. People, like my grandfather, were out of work and burning newspapers to keep their children warm. Little had enough money in hand to keep JAX Lab going through its early start-up days, but the checks he was expecting to carry him into the future were never going to be written. The crash became an abyss. The abyss became the Great Depression. All horizons darkened, especially those in Maine. By November 1929, Little was rationing every cent.

Little suddenly found that he was stranded in Bar Harbor with scientists and staff who were depending on him to save the day. Holed up in a solid brick building constructed to keep feral mice from entering the laboratory and contaminating his stock, Little had to come up with a plan to keep his team and his dream alive. Fortunately, he had one great asset under his roof as disaster settled in—his mice. Little had done an excellent job convincing everyone—the entire country, really—of the incomparable value of his thoroughbreds. As a consequence, the demand for inbred mice had grown nicely. Though many laboratories had learned to breed them on their own, there was the option of selling his to other scientists as a matter of convenience. Little recognized that his way forward through the Great Depression was to become a purveyor of mice, not unlike what Lathrop had done when she was forced to retire from teaching due to her health and her chicken farm going bust. Little spotted an opportunity to breed and sell *superior JAX mice* to other laboratories, advertising them as the best research animals money could buy. He reasoned that if he could fan demand for *his* inbred mice, he'd be able to pick up where he left off before the market crashed and proceed with developing a first-rate cancer research center. He didn't come to this conclusion without a degree of difficulty in recognizing that he would have to postpone the realization of his ambition yet again. But once he came around to seeing his one and only choice, Little concentrated on selling mice and tried to support the research he had begun with a patchwork of seed grants for his scientists and staff. JAX Lab scientists scrambled, too, to secure funding for their work. Though it was a blow for everyone to be thrown into a pit of difficulties, these difficulties were optimistically viewed as temporary. As soon as Roosevelt had sorted out the banks and got the money flowing again, everyone would be back in business, especially the men in Maine who, under the leadership of Little, were busy working hard and weaving laurel wreaths from the green shoots springing up in the field of cancer genetics.

Chapter 5

Bittner

In 1930, after completing his doctorate at the University of Michigan, John Bittner, PhD, followed Clarence Cook Little to Bar Harbor, Maine, to take a position as a staff scientist at JAX Lab. He and other members of Little's team crossbred Lathrop's high-risk mice until they had produced a strain that had an approximately 90 percent lifetime risk of developing breast cancer. One of the goals of studying these mice was to understand the genes responsible for causing breast cancer and to more fully understand the observation that Loeb and Lathrop had made in 1918—that the risk for breast cancer was passed down primarily by females in the high-risk strain. As Loeb and Lathrop had reported, "We may therefore conclude that both mother strain and father strain may prevail; but *the mother strain dominated* in a much larger number of our cases than the father strain."[1]

Loeb and Lathrop had discovered that females from the high-risk strains appeared to impart more than their fair share of breast cancer risk, and Little and his team wanted to confirm that this was, in fact, the case. The fundamental question was whether breast cancer genes were passed down by males and females equally, as Mendel had predicted, or were passed down *unequally*, with the females imparting most of the risk, as Loeb and Lathrop had reported. If the laws of inheritance applied to cancer genes, as they should, then males and females would impart the genes and the risk associated with these genes, equally. If, as Loeb and Lathrop's work suggested, the laws of inheritance did not apply to the genes for breast cancer such that the risk came predominately from the female side of the equation, then what was the source and nature of this apparently female *extrachromosomal factor*?

When Little's team crossbred the high-risk mice with mice from the low-risk strain, they found just what Loeb and Lathrop had found: Mendel's laws did not apply as they do for traits like eye color. They discovered that *only the offspring born of high-risk females* developed breast cancer. If males from the high-risk strain were bred with females from the low-risk strain, the risk for breast cancer did not rise. Again, the high-risk females imparted risk; the high-risk males did not. And yet, both the males and the females from the high-risk strain carried breast cancer genes. These results were perplexing. Those powerful females: What was their secret?

Little's team concurred with Loeb and Lathrop's 1918 report: *it was the females that passed the risk for breast cancer on to offspring*, not the males. Genes certainly played a role in causing breast cancer, but *something else* was swimming around in the gene pool, too, and it seemed to be a decidedly *female influence*. If there was something else besides genes causing breast cancer in inbred mice, what was it? More importantly, exactly where within the females (physically, that is) was it coming from? If it wasn't strictly in the genes—which are carried on the chromosomes—then it must be an extrachromosomal factor. The next step was to discover what that factor was.

Little's team conducted another series of crossbreeding experiments. They took the new strain of *ultra*-high-risk mice—the ones with a 90 percent or so lifetime risk of breast cancer—and crossbred them with Lathrop's original strain of high-risk mice. Again they wondered, would the numbers predicted by Mendel's laws come right if both of these high-risk strains were crossbred? And again, they found that the answer was no. In fact, things got worse! They discovered that the incidence of breast cancer in the offspring produced by these new crossbreeding experiments was even *higher* than it had been in either parent strain.[2] This was very surprising. To understand why the results of this experiment were so baffling, let's return briefly to Mendel's original experiments with plants.

When Mendel crossed green peas with other green peas, he only got more green peas, not *greener* peas. But when Little's team crossed one strain of high-risk mice with another strain of high-risk mice, the offspring developed even *more breast cancer*! Mathematically (in other words, according to the laws of inheritance) this *should not* have occurred. So, either Mendel's Laws were flawed or cancer genes behaved differently than genes for eye or pea color. Or something in addition to cancer genes was playing a role in causing breast cancer. Increasingly, it looked like something in addition to cancer genes was contributing to the formation of breast cancer in the inbred mice. What's more, this extrachromosomal factor was mighty potent. Bittner calculated that it was *six times more powerful* in imparting breast cancer risk than any of the cancer genes involved.

In 1933, Little convened a meeting of the research staff at JAX Lab to develop a strategy for finding the source of the extrachromosomal factor that apparently resided with the females of the strain. Based on the previous work of other scientists, there seemed to be several tissues to consider: ovaries (hormones), blood, placenta, and milk. In later years, Little recalled that Bittner had drawn the lucky straw: he was asked to look for the extrachromosomal factor in the milk. But truth be told, no one else seemed particularly interested in looking in the milk, so Bittner said he would.

Bittner began his investigation by carrying out a series of experiments in which he took newborn mice from the high-risk strain, before they'd had a chance to nurse on the milk from their high-risk mothers, and immediately put them into pens to nurse on mothers from the low-risk strain. He reasoned that if the extrachromosomal factor was *in the breast milk*, then removing infant mice from their high-risk mothers before they had a chance to nurse should lower their subsequent risk for breast cancer. And it did. It lowered the risk from 88 percent to 30 percent. (Another observation that is somewhat tangential, but is nevertheless worth reporting, is that these high-risk strains of mice not only had high rates of breast cancer, they had high rates of lung cancer, too.) Bittner's brilliant foster nursing experiments confirmed that the extrachromosomal factor *resided in the milk of high-risk mice*. There was no question about the importance of this discovery: the experimental results were not just positive, they were overwhelmingly so. Although genes certainly played a role in causing breast cancer in high-risk strains, the *milk influence*, as Bittner called it, was the unrivaled star of the show. Bittner reported his discovery in a preliminary report published in *Science* in 1936, and then moved on quickly to carry out further investigations of the milk factor.[3]

Bittner then conducted the same foster-nursing experiment, but in reverse: he took newborn mice from the low-risk strain and immediately put them into pens to nurse from high-risk mothers. The logic remained the same: if the extrachromosomal factor resided in the milk, then allowing low-risk mice to nurse from high-risk mothers should *increase* their subsequent risk for breast cancer. As expected, exposure to the milk of high-risk mice increased the risk for breast cancer in low-risk mice. In a paper published the following year, Bittner concluded, "These experiments would tend to indicate that some influence is transmitted through the mother's milk which is of prime importance in determining the incidence of breast tumors."[4] Which is to say, Bittner could now say with confidence that cancer genes *and* something in the milk conspired to cause breast cancer in high-risk mice.

Based on all available information, it was increasingly clear that cancer genes, hormonal influences, and the extrachromosomal milk factor were the three ingredients required to create breast cancer in high-risk mice. Now, the next challenge was to identify the relative importance of each factor in producing breast cancer. Bittner expected that these three factors worked together like ingredients in a recipe—proportion was paramount. Bittner had to figure out the proportion of each factor needed to create breast cancer in high-risk mice—especially how much breast milk was required to produce a tumor. What he discovered was that very little breast milk was required: even a few minutes of nursing on milk from high-risk mothers was enough to dramatically increase the subsequent risk for breast cancer.

As soon as Bittner discovered that the female extrachromosomal factor resided in breast milk, and that foster nursing raised or lowered breast cancer risk according to whether infant mice were exposed to the milk factor or shielded from it, he knew, *de facto*, that the milk factor was an infectious agent. There was no other way to understand the results of these foster-nursing experiments except to conclude that *breast cancer in these high-risk mice was, in large part, an infectious disease passed around the litter in the milk of infected mothers.*

Given all the evidence that lay before him, it was logical to hypothesize that *the milk factor was a virus.*

Of course, Bittner would need to replicate his experiments—and other scientists would have to be able to replicate them, too, and get the same results—before he could confidently say that he had discovered a *breast cancer virus*. Nonetheless, Bittner and his colleagues at JAX Lab must have been awed by his discovery, preliminary though it was. If it held up under scrutiny and further experimentation, it would be an enormous breakthrough: a breast cancer virus swimming in the milk!

To go a step further, if a similar breast cancer virus was ever found to exist in women, then it would revolutionize the approach to finding a cure for this disease in people. Infectious diseases caused by viruses can be prevented with the right vaccines. Might this be true for breast cancer? These were nothing more than fleeting thoughts back then, but they must have come to mind. As Sigmund Freud remarked in his book *Interpretation of*

Dreams (1900) about his discovery of the unconscious as manifested in dreams: "Insight such as this falls to one's lot but once in a lifetime."[5]

Bittner had an even more important challenge than making sure that his experimental data were valid and reproducible: the cancer research community maintained an extremely dim view of tumor viruses. Peyton Rous, MD, was the first to propose the existence of a tumor virus in 1911 when he discovered a "cell-free extract" that caused sarcoma (cancer of muscle) in chickens. Very few people believed him. Worse, they ridiculed him. They condemned his experiments. They trashed his results. Quite simply, they refused to accept that tumor viruses existed, and they came down hard on anyone who did. As a result of the scathing skepticism that prevailed, Bittner thought it wise to refer to his discovery as the milk influence both in deference to the preliminary nature of his data and to the pall that the medical establishment had cast upon all tumor virus research.

In December 1938, two years after Bittner discovered the milk agent, the National Advisory Cancer Council published a report, *Fundamental Cancer Research*. The report, which was issued to the medical research community with the gravitas of a stone tablet, stated: "It is considered established that mammalian cancer is not infectious." The council warned the medical community, and anyone else who cared to listen, that all discussion of tumor viruses "could be disregarded."[6] The recommendations of the National Advisory Cancer Council reverberated throughout the research community so effectively that Michael Shimkin, MD, a cancer researcher at the National Cancer Institute, recalled that "cancer research around 1940 was laboring under some premature and unfounded conclusions, one of which was that cancer was not an infection and virus research was a waste of time, despite the examples of the (chicken) sarcomas, mammary tumors in mice, and rabbit papillomas."[7]

Why was it so difficult for otherwise intelligent, capable men to accept the sound, scientific data regarding the existence of tumor viruses emanating from places like Johns Hopkins University, Duke University,

Harvard University, Jackson Memorial Laboratory, and multiple laboratories within the National Institutes of Health and the National Cancer Institute? What, exactly, was the problem? Why were these men going out of their way to oppose tumor virus research? Well, to begin with, the whole idea of tumor viruses didn't make sense given what scientists had long believed to be the inseparable differences between infectious diseases and cancer. It was obvious to everyone, not just doctors, that close contact with cancer patients did not spread the disease to other people the way something like smallpox did. Additionally, infectious diseases tended to course rapidly through a population, producing symptoms (for example, rash, fever, swollen glands, vomiting, diarrhea, and so on) over a relatively short period of time, sometimes in a matter of hours. While the mortality rate might be high in some instances—smallpox, as an example—in most cases of infectious disease, the majority of victims survived and, typically, became immune. Cancer, on the other hand, progressed slowly, inextricably, and fatally in all but the most exceptional cases. It was therefore understandable that the orthodox medical community, including members of the advisory council convened in 1937 to help draw up the charter for the creation of the National Cancer Institute, was convinced that because cancer grows slowly and is not contagious, it was not an infectious disease—and therefore had nothing to do with viruses. Newly emerging scientific evidence, such as that put forth by Rous, Bittner, and others, was disregarded as erroneous or condemned as simply stupid.

It must also be said that when a threatening contradiction emerges from the periphery and begins to touch powerful, pedantic men whose reputations and egos are on the line, their tendency is to dig in and resist the assault rather than lower the drawbridge and invite the revolutionaries into the castle to make their case. In short, the gospel of cancer had long been written and sanctified. It was not open to debate.

With the convergence and strengthening of his early results based on repeat experiments, Bittner came to believe that he had discovered a breast cancer virus in mice. Naturally, he wondered if this or another virus might cause breast cancer in women, too. It would be many, many years before it would be safe to give voice to such thoughts, however. While Bittner continued to work and wait patiently for his data to mount irrefutably in support of the existence of a breast cancer virus, he continued to refer to his discovery as the milk factor, or agent or influence, and, for the most part, steered clear of the implication that what he had discovered was another tumor virus, similar to the one Rous had reported finding in chickens in 1911. He said later, "If I had called it a virus my grant applications would automatically have been put into a category of 'unrespectable proposals.' As long as I used the term *factor* it was respectable genetics."[8]

Bittner continued to experiment with breast cancer–prone mice, but he faced increasing difficulties in finding grant support for his work on the milk factor. If he pitched his investigations in the light of cancer genes, then he had a shot at getting funded by the government. But it was a fine line to walk when where he wanted to go was firmly in the direction of a virus. In 1938, the National Advisory Cancer Council had set the national agenda for cancer research by focusing primarily on the genetic cause of the disease and specifically disregarding all work on tumor viruses.[9] Little's emphasis on using inbred mice to discover cancer genes played perfectly into this national trend—the sales of JAX mice had nearly doubled the previous year. But inadvertently, Little's marketing of inbred mice as the key to finding cancer genes undermined Bittner's pursuit of the milk factor. It helped smooth the waters somewhat when Bittner correctly proposed that three ingredients were necessary for the formation of breast cancer in mice: cancer genes, the milk factor, and female hormones. Still, putting cancer genes at the head of the parade wasn't enough for Bittner to generate stronger support for his work. In 1943, he narrowly escaped complete defunding by the National Cancer Institute when the National Advisory Cancer Council judged his research to be only marginally important. It didn't help matters that the year before, James S. Murphy, MD, of the Rockefeller Institute went out of his way in an article that was published in the *Journal of the American Medical Association* to say, "There is insufficient indication at the present time that viruses play any important role in the general picture (of cancer); therefore, no attempt will be made to discuss at length the possible relation of this group of agents to cancer."[10] (One can imagine Murphy's colleague at the Rockefeller Institute, Dr. Peyton Rous, feeling disheartened by this implicit condemnation of the thirty years Rous had spent researching the chicken sarcoma virus.)

The National Cancer Institute opened its doors in 1937 with the idea of centralizing federally funded cancer research at its facility on the campus of the National Institutes of Health in Bethesda, Maryland. As a consequence, government funding of research at facilities like JAX Lab came under fire. Why duplicate work that could be more efficiently carried out in one location? Some of the scientists who had moved to Bar Harbor to be with Little understood that the funding grid was changing and left JAX Lab for Bethesda or universities where the funding for cancer research tended to be less precarious. Not long after, the National Cancer Institute decided not to fund any more of Bittner's research, leaving him to continue his work with only a bit of money donated from private philanthropy. Meanwhile, he began collaborating with researchers at the University of Minnesota who were interested in his investigations of the extrachromosomal milk factor. Little, who was increasingly consumed with traveling the country to sell JAX mice, asked Bittner to assume an executive role at JAX Lab, which provided another, albeit modest, source of income. Bittner's colleagues were very pleased to see him move into a leadership role.

In an obituary published at the time of his death in 1961, Bittner would be described as a "gifted researcher" of "exceptional brightness" and "inimitable charm."[11] As team leader in the 1930s, Bittner urged the JAX scientists to dig down further on the milk factor. They discovered that in addition to its presence in the milk of high-risk mice, it was also found in the spleen, thymus, and normal breast tissue of the animals. When these tissues were transplanted into the bodies of low-risk mice, the milk factor could be recovered anew. This factor—this breast cancer virus—was to be found throughout the infected animal, and it could be transmitted in more than one way to animals not infected. They then discovered that when the blood of high-risk *males* was injected into otherwise normal mice, the incidence of breast cancer rose significantly. Increasingly, this suggested that the milk factor was an infectious breast cancer virus.

By 1939, the National Cancer Institute's interest in supporting this research at JAX Lab had waned to the point of being nonexistent. Bittner knew that he would have to leave Bar Harbor if he aspired to be anything more than a highly credentialed administrator of a cancer research institute that had been reduced by circumstance to little more than a mouse factory.

Chapter 6

McKnight and Christian

In 1940, the year Bittner became managing director of JAX Lab, he was asked to be the guest speaker at an annual lecture at the University of Minnesota on the topic of cancer research, an invitation that reflected Bittner's growing reputation in the field of cancer genetics. Within two years, he was working collaboratively with scientists there; and by 1943, with his prospects dimming in Bar Harbor, Bittner was ready for a move. As fate would have it, the University of Minnesota was searching for a premier investigator to fill the newly created George Chase Christian Chair for Cancer Research. Carolyn McKnight Christian had endowed the chair in memory of her late husband who died of cancer of the mouth at the age of forty-two. Mrs. Christian was a very wealthy woman. She was the daughter of a prosperous and prominent Minneapolis businessman who'd carved his fortune out of lumber. The land around Minneapolis was deeply forested with trees that were ideally suited to wood construction—a booming industry at the end of the nineteenth century, as America was expanding from one territory to the next. Minneapolis was strategically located near the fall line of the Mississippi River and thus had both hydropower to run lumber mills and a wide, deep waterway to ship the finished wood right down the center of the country and out to every port beyond. Carolyn's father, Sumner T. McKnight, had made a fortune in this business and then quickly parlayed the profits into banking and real estate, creating a substantial empire in the process.

By the time he reached midlife, McKnight had put his imprimatur on Minneapolis's architecture, building one of the city's first skyscrapers, an edifice that was a towering twelve stories high. It was paid for in cash the day it was completed. (When was the last time anyone did *that*?) In addition to plenty of money, McKnight had also earned the respect of his community. A biographical sketch described him as "one of the most sagacious and conservative men in the city."[1] A photograph of McKnight taken in 1895 reveals a supremely confident, handsome man in middle age with an exquisitely turned white mustache and short-cropped beard. He is poised to perfection in an expertly tailored three-piece suit that looks to have been made of the finest wool. McKnight looks rather like a royal understudy for Edward VII, the future king of England.

McKnight had three children: two daughters, Harriett and Carolyn, and a son, Sumner T. McKnight Jr. Carolyn was born in Denver, Colorado, in 1875, and moved with her family two years later to Minneapolis where she lived for the rest of her life. Naturally, she had every advantage that money could buy, including a prestigious education. Carolyn was sent to the Classical School for Girls in Massachusetts, a preparatory academy that had been established as a "farm team" for Smith College, which was conveniently located right across the street. In *A Handbook of Private Schools*, published in 1918, the Classical School for Girls was described as an institution whose mission was to "give a thorough and systematic mental training, and to develop refined and useful womanhood."[2] The student/teacher ratio was sufficiently low so that "careful attention can be given to the work of each girl." As to the curriculum, "A course is given in the study of Architecture with special reference to the English Cathedrals and the French Chateaux. This course is planned for girls looking forward to foreign travel. The Course in Domestic Science includes practice in economical buying, the selection of menus and the arrangement and service of the table, as well as the cooking and serving of food."[3] No doubt, the rules of etiquette were reviewed and reinforced until the faculty was assured they had been mastered before any girl was considered for graduation.

Rather than continuing on at Smith, Carolyn returned home to Minneapolis where she met and married George Chase Christian, a man whose family's history and wealth were a mirror image of her own.

George Chase Christian's father, George H. Christian, was born in Alabama in 1839. As a young man, he went to work in his uncle's store in Albany, New York, and then moved down to New York City, where he worked as a clerk at the Continental Insurance Company. A few years later, Christian left New York City and headed west to Chicago where he worked as a clerk for a flour merchant. It was there that Christian envisioned the fortune that could be made in making cereal. Because bread grew stale so quickly—within a day—it had no shelf life whatsoever. Bread had to be made every day (starting *very early* in the morning, if it was to be ready by breakfast), it had to be sold locally, and it had to be eaten quickly. Cereal, on the other hand, is a dehydrated product made of grain (typically, wheat or rye) that, as a consequence, stays fresh longer. Which is to say, cereal can be made *at leisure*, and it can be *stored*. As Christian saw it, the future lay in finding a way to mass-produce cereal, not bread.

In 1867, Christian moved to Minneapolis, a city that had every ingredient that would be required to make cereal in bulk. Minneapolis was located close to America's "breadbasket," where the wheat used for making flour was grown. The city was strategically located at Saint Anthony Falls on the Mississippi River, where there was a reliable source of hydropower for the mills—lumber and flour alike—and where there was an inexpensive waterway for shipping goods downstream. However, there weren't many flourmills operating in Minneapolis when Christian arrived. The local wheat, which was grown in spring, tasted terrible and spoiled quickly. As a consequence, there was little demand for it and, therefore, little need to mill it. Those who wanted better-quality flour obtained it from businesses that shipped from Baltimore, the nation's flour-milling capital at the time. Christian was convinced that if he could find a way to make Minnesota's spring wheat palatable, he could turn it into a tasty, *storable* cereal and, in the process, make a fortune.

Christian's first move when he arrived in Minneapolis was to get a job as a buyer and manager of a flourmill. Having done that, he was then in a position to learn the intricacies of the business. And while he was learning the ins and outs of flour milling, he made important connections with customers and suppliers, locally and nationally, that would help him launch his cereal empire. In his spare time, Christian began studying how Minnesota's spring wheat might be better milled to make a product competitive on domestic and global markets. He traveled to Europe to learn how the Germans milled their flour, and then he visited France and Switzerland to investigate how milling was done there. After four years of intensive investigation, Christian arrived at a solution for mass-producing Minnesota's spring wheat into a kind of flour that people would actually want to eat. He called his invention the New Process and immediately patented it.

Before Christian's invention of the New Process, flour was milled in relatively small batches. This was a tedious, time-consuming process. It wasn't conducive to mass production, and it certainly would never meet the needs of a growing nation—or fulfill Christian's dream of making boatloads of money. When he arrived in Minneapolis in 1867, the average output of a flourmill was 600 barrels of flour a day. At the time, the population of the United States was thirty-seven million and growing *fast.* He would have needed a loaves and fishes–style miracle to serve that crowd on 600 barrels a day.

Christian's solution was more reliable than a miracle. The New Process sent grain flowing into the milling machines in one continuous stream, like the mighty Mississippi that rolled through the city. Christian's New Process increased mill output 8,000 percent, to 50,000 barrels a day. It wouldn't take many mills to make more money in a day than the US Treasury could print. And most importantly, the flour was delicious! Christian had solved the problem by setting his mind firmly to a solution. There was nothing more to do than to set the wheels in motion and count the money.

Within four years, Christian had made his fortune. He sold his patent for the New Process to General Mills, and then he retired. Christian spent time with his family (two sons and two daughters) and traveled the world. He studied art, science, and philosophy. He spent a great deal of money on philanthropy in Minneapolis, believing that though he had earned every dollar that he made, his fortune was a gift from God that must be shared usefully with as many people as possible. Instead of creating an institution that would memorialize his name, as is the fashion today, Christian chose to create a foundation that reflected his mission to help others: the Citizens' Aid Society. He was especially interested in doing what he could to relieve the terrible suffering of those with tuberculosis, a chronically debilitating disease—often fatal—that affected the poor, huddled masses of the underclass.

George H. Christian sent his son, George Chase Christian, to Phillips Academy, an exclusive high school in Andover, Massachusetts, that served as a preparatory "farm team" for Yale University. Phillips Academy, founded in 1778, was George Washington's first choice for his nephew's education. Clarence Cook Little's great-great-great-great-grandfather, Paul Revere, designed the seal for Phillips Academy, inscribing it with the motto *Finis Origine Pendent* ("The end depends on the beginning"). After completing his studies at Phillips Academy, George Chase Christian chose to attend Harvard rather than Yale. After graduating in 1895, he returned to Minnesota to work in his father's business. In the annual Harvard alumni newsletter the following year, Christian reported, "I have continued in the business of manufacturing flour."[4] Yes, and two years later, on April 27, 1897, he met and married the beautiful Carolyn McKnight, whose family was, as previously discussed, mighty rich, too. The honeymooners traveled to France and England, and then in 1901, they spent the winter in Egypt, which was still one of the many jewels in the British Empire. Upon returning home, George helped with his family's business and also became secretary of his father-in-law's financial institution, the S. T. McKnight Company, which was headquartered in the dazzling, twelve-story skyscraper. In his next submission to *Harvard College Class of 1895*, Christian reported that, in addition to his other responsibilities, he was director of the Bank of Minnesota, director of the Hardwood Manufacturing Company, director of the Midland Linseed Company, and chairman of the Anti-Tuberculosis Committee.

Carolyn McKnight and George Chase Christian were married for twenty-one years, but they never had any children of their own. They did, however, raise four foster children. In 1918, George was diagnosed with inoperable cancer of the mouth. In those days, when a patient was diagnosed with head and neck cancer, there was nothing that could be done except to wait for the dreadful end—and pray that it came quickly and mercifully. Inoperable cancers of the mouth grow extensively in all directions, often to a point where they erode one of the large blood vessels running up and down the neck. This is a catastrophic event that results in massive hemorrhage—a protracted, calamitous conclusion characterized by agony and torment for patient and family alike. The men in the Christian family tended to die in old age of heart disease, so George's ordeal with cancer was as unexpected as it was unwelcome.

George H. Christian died of a heart attack in 1919, the same year his son died of cancer. George's mother had died three years before. With the death of all the direct heirs to the Christian estate, Carolyn McKnight Christian took command of the family's vast and growing fortune. It's no surprise, then, that her philanthropic interests led her beyond tuberculosis and the poor into finding—or making—opportunities to support cancer research. In 1923, she donated $250,000 (approximately $3 million in today's dollars) to finance the construction of the first cancer hospital in the state. The new, fifty-bed Cancer Institute opened at the University of Minnesota in Minneapolis in 1925. Radiation therapy had recently been introduced as a way to stem the growth of solid tumors such as the one that killed her husband. A portion of Mrs. Christian's donation was used to purchase a state-of-the-art X-ray machine. In 1936, she established a scholarship in her husband's name at Harvard University, his alma mater. She also created a memorial lecture in cancer research at the University of Minnesota in honor of her husband. John Bittner was invited to deliver this address in 1940.

In 1943, Mrs. Christian endowed the George Chase Christian Professorship for Cancer Research at the University of Minnesota with the mandate to "understand the cause of cancer" and, thereby, "relieve its suffering." The faculty awarded the chair to John Bittner, and made him director of cancer research. Whether she knew it or not, Mrs. Christian had rescued Bittner's career from grave uncertainty. An indirect result of her philanthropy and the Cancer Institute's choice in offering the newly endowed chair to Bittner was that research on the milk agent was rescued, too.

Even though Bittner had made the move from JAX Lab, where its future as a cancer research center was most uncertain, to an academic center where it was guaranteed, he still faced a life largely spent trying to find grant support for his work and that of the men and women who staffed his lab. This may have been slightly easier coming from a university address, but that didn't make it easy overall.

Time and tide are all: in a peculiar twist of fate and circumstance, Carolyn McKnight Christian provided John Joseph Bittner with an opportunity that became increasingly elusive to his mentor, Clarence Cook Little: a place where he could conduct independent research, following his scientific nose wherever it took him. Of course, Bittner had earned it; but he was lucky to have it.

And we're lucky that he had it, too. Bittner continued investigating the milk agent, and within a few years (thanks to the development of the electron microscope), there was no doubt in anyone's mind, for they could see it with their eyes: the milk agent was a virus: a *breast cancer virus*.

Bittner's youngest daughter, Betsy Bittner Loague, was almost six years old when her family moved from Maine to Minnesota. (Well, at least the winters didn't come as a shock.) As she got older, Betsy accompanied her father on his lecture trips abroad, both as a traveling companion and to help carry his bags. Although he was still relatively young, in his late forties, he suffered from early-onset heart disease. Betsy recalled one trip in particular that stood out in her mind. They were crossing the Atlantic by ship in 1957, headed for Italy, where her father was to receive an honorary doctorate from the University of Perugia. While on board, they ran into Albert Sabin, the man who developed the oral polio vaccine. Bittner and Sabin knew each other, and they palled around during their long, leisurely days at sea. Sabin had received his medical degree at New York University, had worked at the Rockefeller Institute with Peyton Rous, and thereafter became interested in infectious diseases and in vaccines.

In 1952, Jonas Salk, MD, made an injectable polio vaccine that was largely effective in preventing the disease. Recognizing some of the limitations of the Salk vaccine, Sabin began working on an oral preparation—one that would later be known as *the sugar cube*. However, Salk had become something of a demigod, and was a darling of the March of Dimes. The general feeling was that Salk's vaccine needed no improvement and Sabin's efforts were a waste of time. When it became plain that Sabin's vaccine was unwelcome in America, he began looking elsewhere to see if it would work. At the height of the Cold War, this Polish Jew (originally, the family name was Saperstein), whose parents had emigrated in 1921, was on his way to Soviet Russia, to test his vaccine there. According to Betsy Bittner Loague, her father and Sabin commiserated over the trouble they faced in bringing game-changing scientific discoveries to the fore.

Overall, the trip worked out better for Sabin than for Bittner. Between 1956 and 1960, Sabin safely inoculated 100 million Russians, preventing polio in 84 percent of the country's population! Russia forwarded the vaccine to Japan where polio was endemic. In 1957, the US government finally permitted Sabin's oral vaccine to be tested here. In 1963, it became the standard polio vaccine around the world. It was Sabin's oral vaccine that I happily stood in line to receive when I was a little girl—what child didn't want to down a sugar cube at ten? Sabin was ultimately hailed a hero, and in later life began to investigate the link between viruses and cancer, an interest that was sparked during his early years at the Rockefeller Institute and later heightened by the time he spent crossing the ocean with Bittner.

Betsy Bittner Loague posted an entry about her father in Wikipedia several years ago, and she continues to track conversations about him on the Internet. In 2011, she found my blogs about the breast cancer virus on the Breast Health & Healing website, and read with interest my support of furthering this research. She sent me an email, introducing herself to me, and we've kept in touch ever since. Betsy told me that she vividly recalled her first meeting with Carolyn McKnight Christian: "I remember meeting Mrs. Christian. She scared me to death. Her home was massive, in my eyes, overwhelming in grandeur, and I knew I had to be on my best behavior."

Bittner remained at the University of Minnesota for the rest of his life. He continued his research on the milk factor, and within a few years began referring to it as the *mammary tumor agent.* By using the term *mammary tumor agent,* Bittner planted the flag squarely on the proposition that an infectious agent caused breast cancer in mice. This new vocabulary put the hypothesis of a breast cancer virus into a slightly higher gear. As a tenured professor and director of cancer research at a major teaching university, Bittner's work, if still controversial, was more highly esteemed. He was nominated for the Nobel Prize on two occasions, but died too young to be properly recognized for his most important discovery of the milk agent, which was later confirmed to be a breast cancer virus.

Chapter 7

Andervont

Like Clarence Cook Little, the United States Public Health Service can trace its roots to the American Revolution. In 1798, John Adams signed the Act for the Relief of Sick and Disabled Seamen, setting the precedent for government involvement in the management of health and disease. Eighty years later, in 1878, the United States Congress funded what was called the Marine Hospital, located in Washington, DC. Its director was later known as the Surgeon General. That same year, in an effort to prevent contagious diseases from entering the country, Congress created the National Quarantine Act. *Plague houses*, as they were called, were built on Liberty Island, Governors Island, Staten Island, and Ellis Island in New York Harbor as a means of culling sick immigrants before they had a chance to infect others.

In 1912, the Marine Hospital was expanded significantly in response to the increasing needs of the growing American population. Engineers, dentists, scientists, nurses, doctors, and staff were hired in the attempt to control disease and ensure a clean supply of water and sanitary disposal of sewage. This was no small job given the waves of immigrants coming into the country every year, most of whom were very poor and uneducated. The expansion of the Marine Hospital during this period led to the creation of the United States Public Health Service (PHS). Its "officers" wore white uniforms and carried ceremonial swords in keeping with their naval ancestry. The PHS grew steadily over the twentieth century. It was considered a very respectable career for doctors, nurses, and medical researchers, particularly those with an interest in public health. During the Vietnam War, it received an influx of recruits—men who were looking to avoid the draft by serving their country as PHS officers instead.

During its early years, the PHS was more concerned with infectious diseases like polio and tuberculosis than cancer. Though devastating, cancer wasn't yet a common disease, and thus it was not a public health priority. On the other hand, infectious diseases were common and deadly. Typhoid, diphtheria, measles, influenza, and the like could move through a community as efficiently as a John Deere moved through a field of hay. Keeping children alive, keeping mothers alive, controlling the spread of infectious communicable diseases, keeping the water safe for drinking, and keeping sewage under control kept everyone busy. Of necessity, cancer had to wait.

The Public Health Service did not ignore cancer entirely, however. It was considered a most intriguing, if morbid, disease; and, as a matter of curiosity, its research had strong support within the academic community. In 1922, university leaders (under the guidance of the National Advisory Cancer Council—a body of men who set the agenda for cancer research) persuaded Congress to fund two cancer research centers. The centers were to be run under the direction of the PHS, and were collectively known as the Office of Cancer Investigations. One was located in Boston, at Harvard Medical School, and the other center was set up in Washington, DC. Both were serious but low-key operations, at least at first. But as enthusiasm for understanding the genetic cause of cancer mounted—largely due to Little's promotion of inbred mice—the Office of Cancer Investigations at Harvard Medical School got the bug and joined the fray.

J. W. Schereschewsky, MD—fondly and conveniently known as Sherry—was head of the Office of Cancer Investigations at Harvard Medical School and considered his operation the bedrock of all cancer investigations in the country. But unfortunately, Sherry's pursuit of cancer genes had run into a snag. In the spring of 1932, he sent a letter of inquiry—an SOS, really—to Little at JAX Lab. Sherry was having no success inbreeding strains of mice. If he wanted to keep pace with the newly emerging field of cancer genetics, he needed cancer mice. Sherry was stumped and couldn't figure out what, exactly, he was doing wrong: he wasn't breeding thoroughbreds, but duds. Harvard needed help, and thanks to Little's relentless marketing, the whole country knew that he was the man with the mice.

Sherry freely admitted in his letter, "You have colonies in your laboratory in which the spontaneous tumor rate is stabilized."[1] Alas, the mice in his cages were chronically "inadequate."[2] Sherry asked if he could send a member of his team to JAX Lab for a few months to master the art of making inbred mice. Little was delighted to have captured Sherry's attention. A collaboration with Harvard Medical School, the Office of Cancer Investigations, and the Public Health Service would serve Little well. Little told Sherry that he would be more than happy to lend a hand. Not only would he be in a position to demonstrate his skill and expertise, Little would earn the respect and good will of a host of important, powerful men in Boston and Washington, DC.

Little invited Sherry to send an investigator to JAX Lab for the summer, but soon they began to haggle over the cost of the mice that would be used during the summer visit and, later, back at Harvard. The two men finally settled on a price only to get balled up on more fine print: Who should absorb the cost of mice that didn't survive the trip from Maine to Massachusetts? In the end, a deal was struck, and one of Sherry's very best men, Howard Andervont, PhD, moved to Maine for the summer.

Andervont had been working at the Office of Cancer Investigations at Harvard Medical School for several years. He'd graduated from Johns Hopkins University in 1926 with a doctorate in medical science, and had devoted his graduate thesis to studying the relationship between the cow and chicken pox viruses. Andervont was especially interested in the question of whether viruses could jump species, say, from animals to humans. Andervont's interest in the range of infectivity of viruses among different species set the stage for a most interesting career, one that got a whole lot more interesting when he arrived at JAX Lab.

Like just about everyone else involved in cancer research at the time, Sherry's team was focused on using inbred mice to discover the genetic cause of cancer. But Andervont's initial interest, and expertise, was in viruses. He had only come to cancer genetics by virtue of his tenure with Sherry as a member of the PHS at the Office of Cancer Investigations at Harvard Medical School. Andervont was primed to spot and dwell upon any virus that crossed his path. When Little assigned Andervont to work in Bittner's lab for the summer, he would come away with more than just the trick of inbreeding strains of breast cancer mice. Bittner was hot on the trail of the extrachromosomal factor, and there was no better man for Bittner to have at his side than a trained virologist with a mind as clear and brilliant as his own.

Andervont showed Bittner a trick that he'd learned for transplanting tumors into the tails of mice. Tumor transplantation was the subject of Bittner's doctoral thesis, so he greatly appreciated Andervont's suggestion. Bittner, in turn, showed Andervont how to successfully inbreed strains of breast cancer–prone mice. More importantly, the two of them put their heads together over the question of the extrachromosomal factor, the special female-carried ingredient that played a key role in causing cancer.

Meanwhile another problem arose between Little and Sherry. It didn't take long for both men to realize that major discoveries might emerge from the scientific collaboration taking place between Andervont and Bittner around the subject of the extrachromosomal factor. Another round of negotiations ensued. Sherry argued that Andervont had spearheaded the experiments at JAX Lab, even though Bittner had been working on the extrachromosomal factor (using 4,300 mice in his experiments) for more than a year. Little countered that, after all, it was really only a summer's worth of work. Sherry's advantage in this next round of negotiations was the same as in the first: he was the guy with the money—Harvard money, federal money, and plenty of it. Sherry continued to howl that the Office of Cancer Investigations at Harvard Medical School was the bedrock of the national involvement in the problem of cancer, and he had every intention of leveraging every advantage Harvard had to secure a steady stream of federal money to support his research. The PHS was growing. Government involvement in cancer research was growing. JAX Lab was not growing—except by being nice and selling mice.

With the deck so heavily stacked against him, Little had little choice other than to concede to Sherry's claim to the scientific discoveries that might result from the Andervont/Bittner collaboration. Little's capitulation can be viewed in the light of his hope to reconstitute the dream of having a first-rate cancer research center of his own at JAX Lab. He yielded to Sherry because it was politically and financially expedient to do so. No doubt, Little hoped that once the Depression had passed and money was moving freely again, he'd be able to return to his research and fearlessly claim the results of all JAX Lab experiments. Until then he had little choice but to hoist and carry Sherry's flag.

The importance of Andervont's sojourn in Bar Harbor with John Bittner cannot be overstated. He became *deeply* intrigued by the extra-chromosomal factor and stuck with it for the rest of his career. When Andervont returned to Harvard Medical School in the fall of 1932, he wasted no time starting up his own foster-nursing experiments. He replicated and validated Bittner's experiments, and then moved ahead in hot pursuit of the elusive female factor. Andervont and Bittner worked jointly and separately, but on parallel tracks, with inbred mice to identify just what this factor was.

Five years later, in 1937, a powerful group of academic cancer researchers persuaded Congress to establish the National Cancer Institute (NCI).

The NCI was designed to function as a *branch* (not a separate entity) of the National Institutes of Health (NIH) and was to be similarly staffed by members of the Public Health Service Officers. The leadership of the NIH was to oversee the work of the NCI, an organizational arrangement that set the precedent when new institutes were added over the years (for example, the National Institute for Heart, Lung, and Blood in 1948, the National Institute for Allergy and Infectious Disease in 1948, and the National Institute for Mental Health in 1949). This organizational hierarchy was woven into the warp and weft of all of the work that took place under the direction of the NIH over the next thirty-five years.

As we shall see later in the story, when citizen activists who were deeply dissatisfied with the lack of progress being made in curing cancer demanded in 1970 that the National Cancer Institute be torn from the NIH and reconstituted as a separate agency answerable to political appointees, they found, to their chagrin, that a standing army of PHS officers (and their many friends within the academic community) rose up with the passion of starving peasants to protest the change. The activists were stunned when they discovered that these mild, studious men in military dress white uniforms and ceremonial swords constituted an indomitable foe. The newly formed NCI hired scientists, doctors, nurses, and staff as if another war was under way. Andervont was, by then, a top gun investigator at the Office of Cancer Investigations at Harvard Medical School, clearly in the crosshairs of those organizing the NCI in Bethesda, Maryland. Andervont was happy to make the move to the more verdant countryside just outside Washington, DC, and continued to expand on his foster-nursing experiments as soon as he arrived. Bittner published his discovery of the milk agent in 1936, the year before Andervont transferred to the NCI, so Andervont had even more reason to believe that what they'd been looking for was, in fact, a tumor virus.

Back in JAX Lab, Bittner was running low on funds, but not on ideas for new experiments. By 1939, he had formulated a general theory of breast cancer in mice. He hypothesized that three factors had to be present for breast cancer to form:

1. An inherited susceptibility, by which he meant that specific cancer genes conducive to the formation of breast cancer had to be present in the animal
2. Hormonal stimulation, by which he meant secretions made by the ovaries and during pregnancy had to be present (Their absence in males exempted them from breast cancer. Men rarely get breast cancer. Without the ongoing stimulation of the virus provided by estrogen, the breast cancer virus as it exists in males is only rarely able to produce a tumor.)
3. "An active influence in the milk," by which he meant that a factor in the milk had to be present *or the tumors would not form*

An inherited susceptibility (cancer genes) and hormonal stimulation (estrogen and progesterone) were not enough to cause breast cancer in mice. The milk factor *had to be present, too.*

Andervont, who was so busy foster-nursing mice at the NCI that his operation was often referred to as a mouse dairy, began testing Bittner's general theory of breast cancer. In fact, by 1940, the foster-nursing experiments that Bittner pioneered in the early 1930s had become the centerpiece of cancer research at the National Cancer Institute. Andervont replicated Bittner's experiments, using a greater number of mice in his experiments as a way of rendering the results even more robust, and continued to corroborate the data Bittner was generating in his new position at the University of Minnesota. The unspoken conclusion, still in need of verification, was that the mammary tumor agent was a virus—but no one was saying so, not yet.

Andervont enjoyed a wonderful and successful career at the NCI. He eventually rose to the highest position available to a PHS officer: chief of the laboratory of the National Cancer Institute. That is, he was the nation's chief biologist for cancer research. This essentially put Andervont at the head of all the experimental research being conducted by the NCI at its facility in Bethesda and everywhere else in the country where it had funded cancer research. Everyone called him Andy. Not only was he a skillful investigator, he was an adept leader, and playful, too. Weekly poker parties took place in his home every Saturday night. Every New Year's Eve he hosted a party "with piano playing . . . and singing of nostalgic songs, on and off key, and enough alcohol to remember the next day. The usual womanizing, at the level permitted or ignored in those days, of course went on, and a few romances even led to marriages."[3] King Arthur at his Round Table could not have been more pleased.

***

How about this? The first monograph issued by the National Cancer Institute concerned the milk agent. Andervont and his colleague W. Ray Bryan, PhD, a virologist recruited from Duke University, wrote the manuscript and summarized everything that was then known about breast cancer in mice. The monograph, *A Symposium on Mammary Tumors in Mice*, provided scientific evidence that suggested the mammary tumor agent was a virus.[4] They also reported for the first time that immunity to the virus, at least in rabbits, had been demonstrated and confirmed experimentally.

In 1958, Andervont became a trustee of JAX Lab, an honor that brought him, once again, into close contact with Little. In 1961, Andervont became the editor of the *Journal of the National Cancer Institute*, a very prestigious position that he held until he retired seven years later. By the time he retired in 1968, Andervont had studied the milk factor for more than thirty years. Importantly, he confirmed all of Bittner's findings, both at JAX Lab and at the University of Minnesota. And in one seminal experiment, Andervont converted a low-risk strain of mice into a permanently high-risk strain by introducing the milk agent into the population. Once the milk factor had been introduced, it never left the strain. There was no doubt that genes played an important role in the formation of breast cancer in mice; but, again, the milk factor was the key ingredient. Of course, by the mid-1940s, everyone knew that the milk factor (a.k.a. the mammary tumor agent) was a virus. So, for at least the last fifteen years of his tenure at the NCI, Andervont could safely call it what it was: a tumor virus passed in breast milk, *and in seminal fluid*!

In earlier studies carried out at the NCI, Andervont discovered that males shed the breast cancer virus in their seminal fluid. Males from a high-risk strain of inbred mice were able to infect females from a low-risk strain by passing the milk factor to them via sexual intercourse. Apparently, the milk factor was a seminal fluid factor, too. Further evidence suggested that both factors were identical: *a virus*. Andervont found that when high-risk males transmitted the mammary tumor agent to low-risk females via sexual intercourse, the females—who were now infected—did not develop breast cancer, but *their daughters did*. Andervont was the first to report that *breast cancer is a sexually transmitted disease*: males transmit the mammary tumor agent to females, who then pass it on to their daughters. Once introduced, the mammary tumor agent remained present and active in all subsequent generations. Andervont showed that when males transmitted the mammary tumor agent in their seminal fluid, it became an inherited breast cancer agent—like a BRCA mutation, but with a different mechanism of action. (A BRCA mutation is a genetic defect inherited from either the mother or the father that interferes with the normal DNA repair mechanisms in the cell. This defect predisposes the carrier of a BRCA mutation to various types of cancer, with breast cancer being the most common tumor found in women with this mutation. On the other hand, when a female mouse is infected with the mammary tumor virus via the seminal fluid of an infected male, the virus leads to an increased risk for breast cancer by means of directly activating any latent cancer genes that might be present. In later chapters, we'll delve more deeply into the different ways in which tumor viruses and genetic mutations predispose to breast cancer.)

In his years at the NCI, Andervont made another important discovery about the mammary tumor agent that was a real surprise. He discovered that *mice were able to develop passive immunity to the milk agent if they received antiserum from rabbits that were immune to it.* Let me dissect the meaning of this sentence, bit by bit, so that you can understand exactly what I'm talking about and why it may hold great potential for the future of breast cancer in women.

There are two general types of immunity: active and passive. Active immunity occurs when an individual (animal or human) is directly exposed to an infectious agent like, say, the measles virus. Many of the Baby Boomers, including myself, had measles in the 1950s when we were children, long before there was a readily available measles vaccine. Having been directly exposed to the measles virus, we got sick, recovered, and will not get sick again with measles no matter how many times we again bump into the measles virus. We have acquired active immunity to the measles virus because our immune systems actively participated in fighting off the measles virus by developing antibodies to the virus when they saw it, antibodies that eventually eradicated the virus—which is why we recovered. Those antibodies against measles remain with us and protect us from the virus for the rest of our lives. Whenever another measles virus happens to cross our path, our measles antibodies will nail it before it has a chance to do any damage.

Note: On occasion, our immune systems are incapable of making antibodies to a virus. The human immunodeficiency virus, HIV, that causes AIDS, is an example of a virus that our immune systems cannot fight off.

Passive immunity is what takes place when you get a vaccine. For example, children are now routinely given a measles vaccine. The vaccine contains enough elements of the measles virus, but without producing the disease, that it triggers the immune system to make antibodies. Passive immunity by way of vaccination is a clever trick that allows the immune system to manufacture antibodies to a virus without making the patient sick in the process. Vaccinations work incredibly well. Smallpox has been eradicated as a result of the widespread use of the smallpox vaccine. Polio is nearly gone, too, except in poor countries where vaccination is still wanting. The most common cancer in the world is hepatoma (cancer of the liver), and it is entirely preventable with the hepatitis vaccine. Cervical cancer is now completely preventable as a result of the vaccine that prevents infection with the human papillomavirus, HPV. Certain viruses, like influenza, have a propensity to change significantly and rapidly. In such cases, new vaccines must be designed on a regular basis to meet the challenge posed by this type of evolutionary adaptation. Influenza is a virus that is notoriously mutable; and that's why the flu shot must be reformulated and given every year to be effective. Sometimes, the influenza virus changes yet again before the flu shots are distributed, rendering them less effective.

Now, let's return to Andervont's experiment in which he tested for the presence of passive immunity to the mammary tumor agent. He began isolating the mammary tumor agent from mice, and he then injected it into rabbits. Did the rabbits get breast cancer? No. Well, that was interesting. Was it because the mammary tumor agent wasn't able to jump species from mice to rabbits? Andervont wasn't sure. So he took blood from the rabbits that he had injected with the mammary tumor agent and removed all the blood cells. He then took this liquid, or *serum*, and injected it into high-risk mice.

What happened? The mice did not get breast cancer. Why? The rabbits had developed antibodies to the mammary tumor agent, which is why they did not develop breast cancer; and when those antibodies were injected into high-risk mice, they protected the mice from breast cancer, too. The mice did not develop breast cancer because they had been given a kind of passive immunity—antibodies to the mammary tumor agent made by rabbits.

Andervont discovered something else about the mammary tumor agent while he was working at the NCI. He found that when mice were raised in solitary confinement, they developed their breast tumors at an earlier age than mice raised in a community. Apparently, stress—the stress of social isolation—accelerated the action of the mammary tumor agent. Scientists then knew that stress hormones were made in the adrenal glands, so Andervont removed the adrenal glands to see if this would decrease the risk for breast cancer in mice held in solitary confinement. And it did. These experiments proved that, like female sex hormones, the hormones produced by the adrenal glands in response to stress increased the risk for breast cancer, too.

While Bittner and Andervont and everyone else doing research on tumor viruses were trying their best to meet the challenges of breaking into uncharted territory, they had more than simply a lack of knowledge about the terrain standing in their way: The greatest obstacle to progress continued to be the prevailing dogma that cancer was *not* an infectious disease. The accumulating data coming from JAX Lab, the University of Minnesota, and, of all places, the National Cancer Institute was beginning to rattle and challenge the long-held certainty that cancer was not contagious. Unfortunately, the men at the top didn't care: they weren't even listening.

Chapter 8

Gross, Huebner, and Baker

Ludwik Gross, MD, was a general practitioner from a prominent Jewish family in Kraków, Poland, who in 1939 fled Europe barely three steps ahead of the blood and dust of Hitler's tanks. He settled in Cincinnati, Ohio, where he began working at two hospitals—one Jewish, one Catholic. Gross had studied at the Pasteur Institute in Paris, so he was thoroughly familiar with the viruses and bacteria that lay at the heart of infectious disease. When he arrived in Cincinnati, he developed an interest in cancer. Ludwik enlisted in the army when the United States entered World War II, serving until its end. Upon his return, he applied to become an officer of the Public Health Service at the National Cancer Institute, but was turned down. Instead, he joined the staff of the Bronx Veterans Administration Hospital (VA) in New York, where he eventually became chief of cancer research.

Gross was particularly interested in childhood leukemia. He was aware that Rous had discovered a tumor virus in chickens and knew that mice developed leukemia, so he wondered if a virus might cause leukemia, too. There was precious little room at the Bronx VA to house the laboratory mice he needed for his experiments, so Gross kept his mice in the trunk of his car until more space could be found—an action that forever added a touch of geek-in-the-garage mythology to the rest of his storied career.

Gross began by taking serum from mice with leukemia and injecting it into healthy mice, which is essentially what Rous had done with chickens. Nothing happened. Unlike Rous, who was able to grow tumors in healthy chickens by injecting them with serum taken from chickens with sarcoma, Gross was getting nowhere using that trick. But he didn't give up. It would have been perfectly understandable if he had. He might have reasonably concluded that there was no *cell-free agent*—a virus, that is—that caused leukemia in the way the Rous's sarcoma virus caused solid tumors in chickens. Experiments don't lie, and unlike Rous's experiment, Gross's experiment had failed.

Or had it? Experiments don't lie, but neither do they always reveal the truth. Gross's experiment had failed to reveal a leukemia virus, but he was still considering this possibility when one day he was sitting in an audience while a scientist discussed an experiment in which he injected serum into suckling mice. Apparently, using suckling mice had been the secret trick for obtaining the desired result in this lecturer's particular experiment. Gross figured he didn't have anything to lose and he still had plenty of mice teeming in the trunk of his car. He decided to repeat his experiment, but this time he would use suckling mice instead of full-grown mice. He went back to the Bronx VA and injected suckling mice with the serum he had taken from leukemic mice.

Bingo! The mice developed leukemia. More than that, Gross discovered that once his mice developed leukemia, they passed it on to their offspring forever—just as happened to mice exposed to the mammary tumor agent.

Recall from the last chapter that Andrew Andervont (while working first at Harvard Medical School and then at the National Cancer Institute) had created a high-risk strain of breast cancer mice by taking low-risk mice and exposing them to the mammary tumor agent: the mice then developed breast cancer, and all their offspring got breast cancer too. *A new line of high-risk mice had been created by infecting just one mouse.*

Working with mice in the trunk of his car, Gross pulled off the same feat—but with leukemia. Sarcoma, breast cancer, and now leukemia all were evidently transmissible by what looked like nothing less than infectious tumor viruses.

Gross's discovery of a leukemia virus in mice redrew the map for cancer biology in a way that Rous, Bittner, and Andervont's work had not. Perhaps finding a third tumor virus was enough evidence for the existence of tumor viruses that the academic community could no longer simply dismiss them. Or perhaps it was that Gross had nailed an infectious cause of leukemia, a disease that because it primarily affected children (slaughtering them in a matter of weeks) got everyone sitting forward in their chairs and asking, "What?" For whatever reason, the data could no longer be ignored. Not only were tumor viruses powerful agents of malignant transformation—it was now known that they caused sarcoma, breast cancer, and leukemia—but scientists were now certain that once tumor viruses entered a population, they persisted and behaved like an inherited trait generation after generation. Gross's discovery of a mouse leukemia virus, in addition to the ongoing research about other tumor viruses that was being published around the country and around the world, broke the glass—at last.

The peculiar thing about tumor viruses, as compared to other types of viruses, was that tumor viruses could be passed down from mother to child. Viruses like influenza could infect one person after another, *but they never infected unborn children. No child had ever been born with influenza, polio, smallpox, and so on.* So, when scientists found that tumor viruses could infect one animal and thereby become a permanent feature of every subsequent generation, they were baffled, bewildered, perplexed, and hooked. There was no turning back. Now other scientists were willing to venture forth and say that they wanted to know more about these most peculiar, sinister tumor viruses that were slowly but inexorably spreading cancer from one animal to another.

Gross announced that he had found the world's first leukemia virus in 1953. As you can imagine, some of the typical "frank hostility" blew his way. In a paper he published in 1985 about his early years of research, Gross recalled that the scientific community "hesitated strongly to accept *even as a working hypothesis* the existence of a transmissible virus as the cause of tumors."[1] Gross said, "Research projects and experimental approaches radically different from such accepted concepts or theories were not only frowned upon, but often resulted in the refusal of the necessary financial and logistical support needed by the investigator to carry out his proposed studies."[2] But despite the hot resistance, Gross persevered with his investigations.

In addition to his discovery of the mouse leukemia virus, he added to the body of knowledge about the environmental factors that influenced the action of tumor viruses. For example, Gross demonstrated that exposure to radiation could activate a latent tumor virus, rousing it from where it was hiding within the DNA, drawing it out to become a monster malignancy. Gross discovered that certain chemicals could do the same. Today, we call them *carcinogens*. Hormones, stress, radiation, and chemical carcinogens were all added to the growing list of cofactors that influence the action of tumor viruses.

Like Ludwik Gross, Robert Huebner, MD, was an internist by training. He grew up in Cincinnati, Ohio. His father had gone bankrupt running a movie theater during the Great Depression and was forced to spent the rest of his life working as a tailor to support his family. Huebner graduated from public school with average grades, but he excelled when it came to working hard, clocking eighty hours a week as a clerk in a local pharmacy. He attended Xavier University in Cincinnati, a Jesuit college, with initial thoughts of pursuing law. Then he decided on a career in medicine and, in 1938, enrolled in Saint Louis University Medical School.

At the time, medical students were not allowed to work: nothing was to interfere with their studies, especially not a job. But Huebner had no choice; it was work or starve. Routinely, he would be caught—on one occasion he was putting in hours as a bouncer in a brothel—and then he would be suspended from school. After a brief hiatus, he'd be allowed back in and the cycle would begin again.

In 1941, the year before he received his degree, Huebner enlisted as a reservist in the Public Health Service (PHS). World War II was fully under way and he wanted to dodge the bullets, if he could. Every branch in the military was in desperate need of doctors such that freshly minted graduates were drafted into the service before the ink on their degrees was dry. Despite the financial hardships that required him to work clandestinely to support himself through medical school, Huebner graduated at the top of his class. He took his internship in Seattle before moving on to become an active duty officer in the PHS in the US Coast Guard. He was sent to Alaska, where on one occasion he participated in the transport of a shipload of prostitutes back to Seattle. (We can imagine that the nights he spent working as a bouncer in the Cincinnati brothel served him well on board.)

In 1944, Huebner was reassigned to Washington, DC. He took a post at the Public Health Dispensary, found a sweetheart, married her, started a family, and bought a home in Kensington, Maryland, not far from Bethesda, where the National Institutes of Health (NIH) and the National Cancer Institute were growing by the year. While he was working at the Public Health Dispensary in Washington, DC, Huebner became friends with some of the PHS officers from the NIH's Infectious Disease Unit who were assigned there. Huebner immediately caught the bug for infectious disease and made it known that he wouldn't mind transferring to *their* team. Everyone seemed to like him, so they took him on. Huebner described his first day on the job as somewhat disorienting. The supervisor to whom he was assigned escorted him to a completely empty office and left him there without further instructions. Huebner recalled, "Within a few days I had stolen a chair to sit in that I still have twenty-four years later, acquired a desk and some 'loose' laboratory equipment."[3]

At his new job at the Infectious Disease Unit in Bethesda, Huebner was introduced to the world of viral disease, specifically, a group of large viruses known as *rickettsialpox* that live in fleas, ticks, and mites, and cause diseases such as typhus and Rocky Mountain spotted fever. Huebner became involved in tracking down evidence of these viruses in communities around the country by conducting field studies in human populations to determine how these viruses made their way from animals to human beings. Meanwhile, over at the National Cancer Institute, a new crop of young investigators began to be very interested in the subject of tumor viruses. Naturally, the work being done by their colleagues over at the Infectious Disease Unit became a source of increasing interest to these cancer-focused men. It didn't take long before casual conversations that took place at the various institutes and units on the NIH campus led to more formal collaborations between various teams in the search for animal and *human* tumor viruses. No doubt, it helped that the NCI was new, its staff young, and its bureaucracy only loosely woven. These investigators, fresh from school, had the rare privilege of going in the direction of their curiosity, rather than having to adhere to an old and dusty map.

By the time Huebner arrived at the NIH's Infectious Disease Unit, Andrew Andervont and members of his team at the NCI were moving deliberately and methodically into the field of tumor virus research. Andervont and Huebner shared a friend, Wallace P. Rowe, MD, an investigator at the NCI who discovered that some tumor viruses have friends, too: so-called helper viruses that help the primary tumor virus grow a tumor. In another series of experiments, Rowe again demonstrated that once a tumor virus gets into a population of mice, it behaves just like any other gene.

Throughout the 1950s, scientists at the NCI discovered other tumor viruses, which only strengthened their commitment to the field. By virtue of his friendship with Rowe and Andervont, Huebner knew there was increasing buzz on campus around tumor virus research. More than a dozen other investigators (Stewart, Shope, Beard, Kidd, Bryan, Duran-Reynals, Oberlin, Syverton, Friend, Rauscher, Eddy, Ellerman, and Bang), together with Rous, Bittner, and Andervont, were publishing new discoveries about tumor viruses by the month. Huebner steadily moved in the direction of tumor virus research and, as a result of the years he spent looking for rickettsialpox in human populations, began asking the question: Might animal viruses cause cancer in humans?

Over the course of his twenty years investigating outbreaks of viral diseases like Rocky Mountain spotted fever, Huebner had seen with his own eyes that viruses could move easily from animals to humans, where they cause the same or similar diseases (cowpox and smallpox being just two examples, as Andervont had discussed in his doctoral thesis at Johns Hopkins University in 1926). Based on his knowledge of viruses that did jump species, Huebner had an open mind about the possibility that tumor viruses could do the same. More importantly, there was no one at the National Cancer Institute who would get in the way of his holding or pursuing that belief.

After years of experience investigating outbreaks of viral disease in human populations, Huebner had mastered the technique of finding antibodies to the offending viruses in animals and humans. This, in fact, was the evidence that was used to confirm the cause of the disease being investigated. So if there was suspicion of an outbreak of Rocky Mountain spotted fever sweeping through a community, the detection of antibodies to the virus confirmed that it was the culprit responsible for the contagion and not some other infectious agent. Which is to say, if an antibody is present in the blood of an animal or human which is specific to the virus in question, this is proof of an infection with the virus. To put it in the simplest terms, the presence of an antibody equals the presence of the virus. It is not possible to get one without the other.

Since Huebner had spent years tracking down outbreaks of viral diseases in human populations (a field of study known as *epidemiology*), no one knew better what was required to track down human tumor viruses. Furthermore, no one was more strategically placed to lead the search: Huebner's center of operation just happened to be the NCI—the country's cancer research center.

Sarah Stewart, MD, was another one of Huebner's colleagues at the NCI. In 1953, during the course of investigations on the mouse leukemia virus, Stewart had found another tumor virus. This new tumor virus was peculiar to the point of being scary. The mouse *polyomavirus* caused tumors of the breast, respiratory tract, muscle, and bone . . . in mice, rats, rabbits, hamsters, and guinea pigs. This little monster *got around.* Soon, other scientists at the NCI corroborated Stewart's discovery, which put an end to the dogma that had been pouring like acid on tumor virus research that it was all fantastic error—on the order of believing that the Wizard of Oz was real and powerful enough to get you home. No, tumor viruses were the real deal. They could move around the animal kingdom like lions, devouring anything that looked like food. But were they man-eating predators, too?

A huge breakthrough came when scientists in Bethesda learned how to grow viruses in cell culture. (Recall that viruses are obligate intracellular agents: they must first get inside a living cell before they can demonstrate their power.) Once scientists began culturing tumor viruses in living cells and observing them with an electron microscope—watching them enter cells, replicate there, exit, and move on—no one but the willfully blind could continue to trivialize their importance. Indeed, tumor virus fever began to spread throughout the NCI and beyond.

In 1957, scientists at the National Cancer Institute, emboldened by their discoveries of tumor viruses in animals, said "to hell" with all the naysayers and began actively searching for them in humans. Investigators ventured out of their animal-packed laboratories, pushed back from the benches where their tissue cultures were spewing tumor viruses, and walked boldly into the NCI's cancer wards, where children with leukemia lined the wards in small white beds awaiting certain death. They searched for *and found* viral "particles" in children with leukemia, viral particles that appeared to be identical to the mouse leukemia virus that Gross had discovered several years before.

The first viral particles found in children with leukemia were discovered in their lymph nodes. During the 1950s, 70 percent of children diagnosed with leukemia were dead within two months. Within four months, they were all dead. With the discovery of leukemia virus particles in the lymph nodes of children with leukemia, scientists wanted to be sure that they weren't "seeing things." And so, they examined the blood of eight children with leukemia. Again they found the virus there. To make sure that they had not inadvertently contaminated their tissue specimens (for example, the lymph nodes or the vials of blood) with mouse leukemia viruses that might have "blown in" on their lab coats or equipment, the investigators searched for evidence of the leukemia virus in the blood of thirty-six healthy children. They found no leukemia viruses in healthy children; they only found them in children with leukemia.

As you can imagine, the discovery of what looked like mouse leukemia viruses in children with leukemia was disturbing, enlightening, and transforming. Eight children who were lying on their deathbeds on the pediatric ward at the National Cancer Institute were found to have had viral particles that appeared to be identical to viruses that scientists *knew* caused leukemia in mice and other animals. Healthy children did not have these viral particles. As former director of the National Cancer Institute, Carl Baker, MD, recalled in later years, "This finding led to the possibility that viruses caused human leukemia and perhaps other human tumors and that vaccines capable of preventing such cancers could be produced."[4]

Indeed, the first human leukemia virus, HTLV-1, was discovered in 1979. Millions of people around the world are infected with this virus, most significantly in Japan. Every year, 500,000 people who are infected with HTLV-1 develop acute leukemia, with the vast majority dying in a very short period of time. It is a rapidly fatal disease for which there is no cure and, as yet, no vaccine. The recent implementation of a screening blood test to identify Japanese women who are infected with HTLV-1 has led to a decrease in the incidence of acute leukemia in that country. The virus is passed in breast milk and infected women in Japan are advised to avoid breastfeeding, which has led to a decrease in the number of reported cases.[5]

A couple of things happened just about the time of Dr. Stewart's finding at the NCI, which culminated on an entirely new mindset and commitment to research on tumor viruses. The first was that Jonas Salk had created the first polio vaccine, and he began to openly discuss the possibility of developing vaccines designed to target tumor viruses—anti-cancer vaccines. With the likelihood of eradicating polio thanks to Salk's vaccine, the hope of eradicating cancer followed in its wake. The discovery in 1957 of the mouse leukemia virus in children with leukemia certainly fanned that hope. Wendell Stanley, MD, director of the National Cancer Institute, stood on solid ground the following year when he appeared before the Congressional Appropriations Committee in 1958 to "emphasize progress in cancer virology."[6] Armed with the conviction that scientific evidence provides, he asked for a special appropriation so that the NCI could move ahead as quickly as possible in their investigations of tumor viruses.

Carl Baker, MD, who worked at the NCI for decades and was its director from 1970 to 1972, said that Stanley "called for expanded research in viruses and cancer work and presented scientific evidence supporting the call for the expansion."[7] Stanley made it clear, once and for all: "Viruses were shown to be involved in the induction of cancers."[8] To thwart any skeptics listening in he said, "The notion that viruses could cause cancer in man was of growing acceptance among cancer investigators."[9]

Thanks to Stanley, Congress caught tumor virus fever, too, and called for "a vigorous effort to stimulate research and training efforts in the study of the possible viral origin of human cancers."[10] The appropriation was approved and a check was issued for $1 million for virus and cancer efforts. This would seem like small change now, but at the time it was a significant sum of money. To put this in context, the entire operating budget for the National Cancer Institute was just $27 million. The following year, Congress doubled down and gave the NCI another $2 million for tumor virus research. An equivalent sum, as a percentage of the NCI's operating budget for 2017, would be $252 million—real money.

The leaders of the NCI immediately convened a meeting to figure out how best to spend the money. They didn't want to spray the funds around like buckshot, but wanted to design a coordinated effort aimed at producing the most useful results. To this end, the NCI invited a group of scientific experts whose mission was to draw up a strategic plan for rolling out a new research program (commonly referred to as a "study section") to study tumor viruses. The advisory group was named the Panel on Viruses and Cancer, and its focus was to search for human tumor–causing viruses and determine if vaccines could be designed to prevent the cancers they caused. While the panel was drafting its strategic plan for the new study section on tumor viruses, the NCI held an invitation-only meeting of tumor virus experts to discuss "The Role of Viruses in Relation to Human Malignancies." Drs. Jonas Salk and Albert Sabin were the most notable celebrities in attendance, giving further credence to the subject under discussion. At the conclusion of these preliminary meetings, the consensus was, "After thorough consideration, Study Section expresses unanimous approval of the principle of all possible support of this field."[11]

With the endorsement of tumor virus research by the NCI followed by solid funding of it by Congress, other players began pouring onto the field. The following year (1959), the American Cancer Society sponsored a three-day symposium, "The Possible Role of Viruses in Cancer," converting the once-scorned tumor viruses into little cancer celebrities. Twelve months later, the Congressional Appropriations Committee again increased the NIH's budget to study tumor viruses, agreeing with the NIH director that this research was worth supporting and expanding.

With money, and men, and an agenda, it was time for the NCI to begin tumor virus research in earnest. In 1962, it held its first meeting of the Human Cancer Virus Task Force. The objective was to draw up a five-year plan for identifying human tumor viruses and, to the extent possible, begin development of preventive cancer vaccines. The mammary tumor virus; solid tumor viruses, such as the polyomavirus; and the leukemia virus topped the list.

Money kept pouring in from Congress. In 1963, the NCI received $4.7 million for the study section on human cancer viruses. (Chemotherapy research, which looked increasingly promising, received $24.6 million that year.)

In 1964, the NCI appealed again to Congress: evidence now suggested that an agent responsible for causing human leukemia could be isolated. The NCI was optimistic, saying, "We might be able to prevent this lethal disease."[12] The Appropriations Committee gave the NCI another $10 million specifically to expand leukemia virus research. These were very exciting and prosperous times for science. The nation was headed for the moon—John Glenn had orbited the Earth, and now NASA was sending two men up at a time. If we were conquering space, one mission at a time, then it seemed reasonable that we could conquer cancer, too. And if tumor viruses were the only things standing in the way, well, then we'd develop preventive vaccines for them, just as we'd done for polio. Nothing seemed outrageous in thinking expansively in these terms.

Robert Huebner's work had, by that time, led him to the conclusion that tumor viruses were capable of hiding out in the DNA like stowaway genes until they were summoned forth by carcinogens to produce malignant tumors. He said, "Viewed in this light, the application of radiation, chemical carcinogens, and the natural aging process are believed to 'switch on' the viral genome (in other words, virus genes)."[13] Eventually, Huebner's research, coupled with similar investigations undertaken by other scientists at the NCI (Harold Varmus and J. Michael Bishop) led to the discovery of the first *oncogene*, a type of gene that acts like switches and controls cell growth. When an oncogene malfunctions, it leads to abnormal growth and paves the way for cancer. According to Carl Baker's memoir (see Bibliography), by 1965, as we were getting closer to the moon, the NCI's investigation of tumor viruses had definitely "moved into high gear."[14]

In 1966, five years after Bittner's early death from heart disease, the National Cancer Institute began to strongly get behind research on his mammary tumor agent, which scientists knew by then was a virus: the mouse mammary tumor virus, MMTV. Congress gave the NCI another shot of money—$1.6 million, specifically to intensify efforts in breast cancer research. In 1967, the NCI's budget for studying tumor viruses was increased to $19.2 million.

In 1970, the National Cancer Institute, whose forty-eight standing advisory committees scrutinized every aspect of research and vetted every official statement before it was released to the public, issued a summary report on what it was then calling its Viral Carcinogenesis Branch. The authors of the report formally introduced a new hypothesis about how and why cancers grow: viruses introduce cancer genes into the cells of animals and humans. These viral cancer genes become a part of the rest of the cellular DNA. Afterward, the viral cancer genes are passed down generation after generation as part of the pattern of normal genetic inheritance. The report, issued forty-four years ago, hypothesized that tumor viruses were a general cause of cancer and that "spontaneous cancer as well as cancer evoked by exogenous environmental pollutants, endogenous physiological aberrations, genetic defects, as well as cancer clearly induced by viruses in experimental animals" might pave the way for understanding the cause of cancer in man.[15]

There was another important item in the NCI's 1970 report about its tumor virus research program, and it has far-reaching consequences for our present understanding of viruses and cancer:

> Thus, the concept of built-in virogenes and oncogenes, the expression of which are recognized by both the whole organism and the individual cells as "self" very likely explain the failure of current approaches to control the majority of cancers. Contemporary therapeutic efforts based on destruction of tumors, surgical and radiation treatments and experimental immunological vaccines might now be viewed as largely palliative and, more frequently than not, temporary in their effects on cancer.[16]

Malfunctioning oncogenes and viral cancer genes turn the cell into a malignant renegade, but because these changes take place deep inside the cell, the immune system is tricked into thinking that all is well—at least on the surface. The NCI's prediction in 1970 that cancer treatments, no matter how innovative, are more palliative than curative would prove prescient forty-five years later, in 2015, when Ken Burns produced his documentary film *The Emperor of All Maladies* just as I was putting the finishing touches on this book.

Robert Huebner, MD, was one of the authors of the NCI's 1970 report about its tumor virus research. As a man who'd begun his career looking for ways to control infectious diseases in the community, he argued that prevention was the best means for controlling cancer. Of course, treatment was paramount for those who had the disease. But

Andervont felt that public health policy had to focus its greatest efforts on discovery of the causes of cancer: the tumor viruses. Once these were identified and understood, preventive strategies, such as vaccines, removal of chemical carcinogens, and so on, would then have a greater impact than a world of treatments could provide. Huebner's point was that prevention was the best medicine for the *majority of the population who are not sick, but are vulnerable.* Huebner took the opportunity to remind the public, Congress, and the forty-eight advisory committees whose job was to oversee what the National Cancer Institute was doing with all of the tax revenue flowing to it from Capitol Hill that *prevention was every bit as important for controlling cancer as it was for controlling other forms of infectious disease.*

The National Cancer Institute issues a report every year called the *Annual Program Review.* In 1970, the report stated that viral particles had been seen by electron microscopes in breast tumors taken from rats, monkeys, and humans. The report described finding viral particles in cell cultures grown from these malignant tumors. Which is to say, scientists at the NCI had found tumor viruses in human breast cancers, had taken some of those cancer cells and grown them in culture, and then had found the same viruses growing there. The milk agent that Bittner had discovered in 1936, and that had been shown by electron microscopy to be a breast cancer virus, was being found by NCI scientists in rats, monkeys, and women. Scientists at other cancer research centers, such as James Holland, MD, of Roswell Park, were beginning to find and report the same thing.

The National Advisory Cancer Council (NACC), established by Congress to oversee the NCI, issued an annual report, too. It was one of the forty-eight committees overseeing the work of the NCI. Indeed, it was the most powerful of all, for it was stacked with the most influential men from academia and corporate America. The NACC's primary role was to review the NCI's progress and make specific recommendations for future research. The following passage taken from the NACC's Annual Report in 1971 discusses what was known about the breast cancer virus. Even though portions of the excerpt contain scientific language that might be unfamiliar to the general reader, I've chosen to present the passage in its entirety to emphasize the extent to which the members of the National Advisory Cancer Council endorsed the hunt for a human breast cancer virus and were enthused by the NCI's success in finding it.

> In the Special Virus Cancer Program, because of new research leads and development of new techniques, a new segment of breast cancer was added. . . . The potential existence in humans of an infectious breast cancer virus similar to that of mice, together with epidemiological evidence of "clustering" of breast cancer in some human families similar to that observed in the earliest studies of cancer in mice, led to systematic viral studies on this human disease. Particles resembling the Type B virus of mouse breast cancer were observed in 40 percent or more of milk specimens from women with breast cancer, as well as from healthy women of high-risk populations (high breast cancer families, inbred Parsi sect of Bombay, India) as compared with a frequency of only about 6 percent for specimens from healthy women of the general population. Similar particles were also observed in two tissue culture lines of human breast cancer that had been successfully grown in the laboratory. Moreover, Spiegelman and Moore and colleagues demonstrated the presence of appropriate RNA and reverse transcriptase in purified viral particles isolated from human milk. These studies were extended to examine human breast tumors. Nineteen of twenty-nine specimens of human breast cancers yielded microsomal fractions that hybridized with DNA complementary for mouse MTV RNA. Normal breast tissue specimens or breast tissue from various benign lesions gave microsomal RNA fractions that did not hybridize with DNA complementary to MTV. The above findings suggest that human breast tumors contain functional

> genes that are related to the genes contained in the virus known to induce mammary tumor in mice.[17]

The NACC also reported that more than a hundred cancer-causing viruses had been identified in animals.

In 1971, at the Thirteenth Annual Seminar for Science Writers sponsored by the American Cancer Society, Carl Baker, then director of the National Cancer Institute, reported, "The current status of viral oncology was a major topic at the meeting."[18] Baker discussed seven objectives formulated by the NCI that were aimed at addressing the country's growing cancer problem. The first five of these focused on *causation* and *prevention of cancer*. Baker told the assembled audience of writers, who he knew would be reporting his remarks in all the major newspapers and magazines, that the NCI's goal was to identify tumor viruses, prove that they caused cancer in humans, and develop preventive vaccines. Treatment of cancer appeared farther down the list, coming in as objective number six: "To cure as many patients as possible, and to maintain maximum control of the cancerous process in patients not cured."[19]

Objective number seven was "To restore patients with residual deficits as a consequence of their disease or treatment to as nearly a normal functioning state as possible."[20] As far as Baker and the NCI were concerned—and with the support of the NACC—the race was not to find a cure, but to *find the cause.*

But despite the consensus opinion in favor of prevention that Baker expressed that day, catastrophe was under way. *Catastrophe* is a word whose Greek root means "to overturn," as when a cart topples and spills its contents. The word implies an upheaval that results in ruin. No words more aptly describe what was taking place around the National Cancer Institute than *upheaval*, *ruin*, and *catastrophe*—for tumor virus research and cancer prevention. By the end of the year, the first five objectives were effectively scratched off the list. Objective number six—treatment with intent to cure—became the focus.

Never mind what the NCI thought was best: finding a cure for cancer; a concept marketed more for its appeal than its feasibility became instead a national preoccupation. President Richard Nixon promised to cure cancer. He said so in his inaugural address. He promised to cure cancer in five years, by 1976, in time for the nation's bicentennial. Nixon's promise, fueled by the rabid passion of lay activists who thought cancer curable by fiat, set off a firestorm of diagnosis and treatment and races for cures. Politically and financially, the War on Cancer greased every wheel in sight save for the ones that had been turning for years around the hunt for tumor viruses.

At the helm of this catastrophe for primary prevention of cancer was a woman with a degree in art history who thought that basic medical research was a waste of time and that doctors ought to focus more attention on how to cure their patients. With the help of her friends and vast sums of money directed toward high-powered lobbying, she pushed basic medical research and its corollary, prevention, aside and drove the country headlong in the direction of treatments and cures. Her name was Mary Lasker.

Chapter 9

Lasker

Rumor has it that Edmund Ruffin, a Confederate soldier from Virginia serving in South Carolina, fired the shot that started the Civil War. The Southern troops assembled along the banks of Charleston Harbor that day (April 12, 1861) allowed Ruffin the honor, the story goes, because everyone agreed that he'd started the whole thing anyway. One hundred years later and farther north in Washington, DC, it can be said with more truth than myth that Mary Lasker, standing on a parapet she'd built brick by golden brick in the nation's capital, fired the sniper's shot that started the War on Cancer.

Lasker's maiden name was Woodard. She was born in Watertown, Wisconsin, in 1900. Her father was president of the Watertown Bank. Her mother was an industrious, self-sufficient Irish immigrant who became the most successful sales clerk in a Chicago department store before she retired into marriage in her late thirties. Mary's mother was thirty-nine when Mary was born, which is not uncommon now, but at the time it was almost unheard of. A sister was born five years later, even more remarkably.

Mary was a sickly child, chronically plagued by ear infections. The first antibiotic wasn't discovered until 1928, so very little could be done during Mary's childhood to relieve her suffering or hasten her recovery. The pain, so exquisitely felt in so tender a place (the eardrum), was excruciating—akin to having nails hammered into your skull. Calling the doctor only made things worse—at least, as far as Mary was concerned. Adults would be summoned from around the house and gather ominously in her room, joining the doctor. Already fevered and agitated, as everyone flanked her bed, Mary would become hysterical and need to be restrained while the doctor jabbed something sharp into her bulging ear to relieve the pressure and drain the pus. The ordeal, repeated year in and year out, was more appropriate to a dungeon than a nursery. Mary's anguished screams must have sent shock waves through the house and down the street. If the doctor was not sent for and instead the illness was left to run its course, the eardrum would stretch beyond its ability to endure and would rupture on its own, raising the age-old question of whether the doctor had done anybody any good.

Mary also suffered with chronic bouts of dysentery. Rehydration with intravenous fluids was not routinely done back then, and so these episodes were every bit as debilitating as her ear infections. Mary's earliest childhood memories were of being painfully sick and wrung out, physically and emotionally. However, once she'd recovered from her illness, it was clear that Mary had a ferocious fight in her. Her strength returned with compounding interest when it came to her will to live. Naturally, her anxious, aging parents fretted and overindulged their firstborn child when she was ill. This only made Mary peevish, demanding, and self-centered when she was well.

Mary's traumas made her extremely wary when it came to doctors, which is understandable in a child. But she remained jaded as an adult. The entire medical profession became a voodoo doll for her whenever she felt the need to vent her fury about disease and illness, which was often. It didn't help assuage her indignation when doctors condescended to her, which was often. This only fanned the flames of rage. Rather than develop a mature appreciation for the perpetual challenges of medicine or a wiser response to the vanities of men, Mary's opinion of doctors remained petrified at the level of a furious child. She firmly believed that doctors were willfully ignorant, callous, and directly responsible for the suffering of their patients—in every case. By her own admission, she remained "deeply resentful"[1] of the medical profession for the remainder of her life. It was "just a matter of human ignorance and oversight and lack of intelligent study and training and research. I was infuriated," she said. Doctors weren't interested in improving their skills: "The concept that medical research paid off in lives and in dollars was so unfamiliar to most of these people that they just resisted it." While Mary's anger can be traced to the bewildered rage she'd felt as a child, she took pride in remaining steadfastly inconsolable for the rest of her life. When, in middle age, she reflected on the path she had taken to become a crusader for medical research in a history she recorded for Columbia University's Notable New Yorkers series, she confessed that her "major motivations in doing anything . . . all went back to my violent reaction and hostility to illness."

Mary described herself as a precocious child. She said that she found school tedious because "it was very dull to wait for others to get along with their lessons," explaining, "I was always too old for my contemporaries." She graduated high school in 1918, and then enrolled at the University of Wisconsin, where she barely made it through her freshman year. Mary was a party girl. She plowed through twenty-five boyfriends that she could recall in the course of two semesters. She was less enthusiastic about her schoolwork because she found it boring. Mary returned home the following spring physically exhausted and academically inert. Her mother, Sara, was waiting for her with hooded eyes. Sara was never provided the luxury of a college education and she was not about to let her eldest daughter squander an opportunity that she herself longed for, but would never have. The decision was made to pack Mary off in another direction, Radcliffe College (the college for women at Harvard University), where it was hoped she would settle down and get serious about her education. Mary demanded that her father buy her a fur coat before she left because she got very cold in winter and she "needed it." Her father was appalled at the extravagance that she cloaked as a necessity. But he gave in.

Mary thrived at Radcliffe. As she recalled in her oral history, the coursework was more demanding, but also more to her liking. She developed a passion for art although she had no talent for producing it. She did, however, possess a clever, insightful eye. Mary graduated Radcliffe in 1923 with a bachelor's degree in art history and went off to Europe for what was known then as a grand tour. At the end of her travels to all the major cities and museums, she remained in England for the summer to study Greek sculpture at Oxford University.

Mary returned home to Watertown, Wisconsin, in the fall, but that was only to pick up her things and leave again for good. Her explanation was, "When I graduated college it was an era when no respectable girl would ever think of going home to a small town in the Middlewest. It was always absolutely required that they work in New York for a certain time at least, and I considered it absolutely too dull for words to return to Watertown, Wisconsin. Absolutely impossible." Oh, and there were "certainly no men in that area that would be interesting."

Mary left Watertown forever and moved to New York City, where she shared an apartment with a smart set of equally ambitious young women. The Roaring Twenties were fully under way. Women were drinking and smoking and jumping up and down to jazz. Their hemlines were up off the ground, enough to see some leg and imagine more. Corsets were gone. Underwear was thin and scant and made of silk. Dresses hung from the shoulders, allowing the body to move freely and protrude where it may. Soon after arriving in New York, Mary found an entry-level position at a small art gallery. After staying there for a short while, she angled her Harvard pedigree to catch a bigger fish, snagging a job at Reinhardt Gallery, a high-end purveyor of old European masters.

Mary loved the Rubenses and the Titians, but she also developed an eye for modern art. She understood its power to move the viewer more deeply, beyond admiration for the skill of representation to the core of human experience where feelings are raw and inexpressible. She knew that it was just a matter of time before the modern art movement in Europe would make its way to America, where it would find expression and a ready audience waiting for it. Mary asked Paul Reinhardt, the gallery's owner, to allow her to organize America's first exhibition of Marc Chagall. It was an impressively prescient—precocious?—move for someone so inexperienced in the field. Paul was easily persuaded. He had fallen in love with her. But Paul Reinhardt was not a man that her parents would have wished for her. He was twelve years older, a widower, and an alcoholic. She couldn't do a thing about the first two attributes, but she was determined to iron out the third. Mary demanded that Paul stop drinking. He did—for a year. Satisfied with what she thought was victory, she married him in 1926.

As many a woman before and since, Mary relied on her sweetheart's sweet promises only to discover that they were built on the fog of promises renewed. When the stock market crashed in 1929, Paul crashed too. He took to the bottle and began to drink himself to death. "I was in despair and I couldn't get him to stop drinking. I couldn't get him to do anything," she lamented. Mary left the gallery the following year, went to Reno, and got a divorce. When she returned to Manhattan six months later (she was thirty-four years old), she was in a tight spot: unemployed, alone, aggrieved at the dashing of her dreams. In many ways, she was in the same tight spot as Clarence Little and the rest of the country, searching for a way to pay the bills. But behind the pretty camouflage of tailored clothes and a Harvard degree stood a formidable woman who was highly intelligent, resolved, undaunted. She was not going to survive, she was going to succeed.

As respite during her great upheaval, Mary dove into movie houses for cover and release. She became enthralled with Hollywood and began to wonder if a first-rate movie magazine might be a hit. Her friend Mary

McSweeney was married to a man who worked for the publisher Condé Nast. When she discussed her idea with him, he suggested that she do something with fashion and Hollywood instead. And so, on March 6, 1933, the day President Franklin Roosevelt closed the banks for a week, the two Marys launched Hollywood Patterns, sewing patterns with an aura of the silver screen.

Most American women made at least some, if not all, of their clothing. You'd have to look hard to find a home without a sewing machine, typically a Singer, parked somewhere in a bedroom or in the basement. This homemaking skill had become even more of a necessity during the Depression. By the time Hollywood Patterns hit the stores, one hundred thousand workers had been losing their jobs *every single day* since the stock market crashed in October 1929. Roosevelt closed the banks to try to resuscitate a moribund economy, but still there was no end in sight. Almost a third of America's workforce stood hopelessly in line, waiting for a cup of thin soup or a slice of bread. (My grandfather, who lost the two homes he owned, scavenged for newspapers, tied them into knots, and burned them in the fireplace of a house to keep the family warm. My grandmother pawned her jewelry. Only my great-aunt's grand piano survived, probably because no one wanted it. My grandmother's jewelry was eventually rescued. The family got back on its feet. The old piano was refurbished several months ago and now lives with my sister in the Shenandoah Valley.)

Despite the Depression and the gloom, the two Marys pressed on with their enterprise, marketing dress patterns that featured a different Hollywood star on each envelope. Hollywood Patterns were sold in dime stores like Woolworth's—chain stores that Mary Lasker had never heard of before starting the business. With their fingers placed securely on the pulse of female fantasies, the two Marys marketed Hollywood Patterns to average women as the "chance to make dresses like those the movie stars wear." The business secured releases from the actresses that were featured on the envelopes of their patterns, but it did not pay the actresses a royalty. "No, nothing at all, they just got publicity," Mary said. Free publicity for movie stars, fantasies for women in reduced circumstances, and a market as wide as the country was big turned Hollywood Fashions into a very profitable business.

Mary was surprised at her success, or so she said. At the depth of the Great Depression, she was clearing $25,000 a year. It was enough to pay for a penthouse on East 52nd Street in Manhattan. With money to spare and nothing to tie her down, Mary began to let off steam productively: she put on the cloak of philanthropy and set out to change the world. This was a votive task, one that she had learned at her mother's knee. Sara Woodard had always been active in her community, building parks, seeing to the care of patients with tuberculosis, and working for clean air. Like all children, Mary learned by observation and imitation. She didn't have far to look to find a worthy cause: birth control became her first crusade. Mary was introduced to Margaret Sanger, the nurse who opened the first birth control clinic for women in New York City in 1916, and who had founded the American Birth Control League in 1921. Mary joined the league and fought alongside Sanger to liberate women from unplanned, unwanted pregnancies. This was, and forever remained, her most ardent philanthropic passion. She declared late in life that a woman's right to control the number of children she had was "the most important cause there is."

Interestingly, though married when still young and (presumably) fertile, Mary never had any children of her own. Nevertheless, birth control must have been an important consideration for her. She joined the board of the American Birth Control League. In 1938, she became its president. This was a position that Clarence Cook Little once also held.

Despite the most unlikely and difficult circumstances, Mary succeeded in becoming a shrewd, capable, financially independent New York businesswoman and socialite. Her talent for charming friends and making new acquaintances grew more skillful by the year. By the time she turned forty, Mary was an entrepreneur, an art collector, a hostess, a *women's libber* (before the term was coined), a philanthropist, and a leader. But although her accomplishments were varied and impressive, these were only softly arching warm-up pitches. Her spitballs and fastballs would follow later, once she found a more serious game to play.

On April 1, 1939, Mary was having lunch with friends at the tony 21 Club in New York City when a strikingly handsome, older gentleman whom she recognized brushed by her table on his way out the door to take a telephone call. The dashing blade was Albert Lasker, a German-born advertising genius who had made a fortune in advertising by featuring logos and establishing brands. ("Lucky Strike means fine tobacco," "Double your pleasure with Doublemint gum," Sunkist oranges, Sun Maid raisins, Kotex, Kleenex, Pepsodent toothpaste, and Frigidaire, to name a few.) Mary was astonished when "he passed by my table and didn't look at me, and I thought, 'That man is making a great mistake not to pay attention to me.'" With his white hair and dark eyes, she found him "very, very arresting looking." The following evening Mary's friends invited her to dinner in their home so that the two could meet and get to know each other. Unfortunately, Mary misunderstood the invitation, thinking she was going to a cocktail party and that she could show up whenever she got around to it. Mary arrived an hour late only to find that Albert (who was punctual to a fault) was "extremely cross." She was stunned: "I didn't expect anyone to be cross at me about anything." It didn't take long for all to be forgiven, and everyone enjoyed a lovely meal.

Mary was thirty-nine. Albert was sixty years old. His wife of thirty-five years had died not long before and he had subsequently rushed to marry Doris Kenyon, a forgettable pixie actress who'd once starred opposite icons like Rudolph Valentino and John Barrymore. This hasty second marriage quickly collapsed into a bitter divorce that was coming to an end the day Albert brushed by Mary's table at the 21 Club. Despite their unsettled start and all the other obstacles that might have gotten in their way, Albert and Mary were immediately and strongly attracted to each other. Both were highly intelligent and extremely demanding. Both loved art, architecture, gardens, and travel. They were both animated, sparkling conversationalists who moved in powerful circles and loved every minute of it. Financially, they were independent and astute. "He had an extraordinary quality of vitality," Mary said. "Altogether he was the most brilliant man I ever met."

It's not known what Albert thought of Mary, but he tracked her down and began calling on her regularly. Two months later, on June 21, 1939, Mary invited him to a party on the terrace of her penthouse. They were married exactly one year later. Mary's mother was "absolutely shocked" because Albert was old enough to be her father, but Mary said, "I was determined to see him because I thought he was the most interesting man I knew." Mary and Albert decided to keep their wedding absolutely private, otherwise they would have had to host a grand affair that neither one cared to endure. Instead, they exchanged their vows in front of a judge at city hall in a ceremony that cost two dollars. They took a short honeymoon on a yacht, and then headed to the 1940 Republican Convention because Albert was a delegate from Illinois that year.

Albert treated Mary as if she were a queen. He ignored her when she told him that she wanted to pay for her share of their living expenses. (At the time, Albert owned a fabulous home on 350 acres in Lake Forest, Illinois. It had pools, a movie theater, spectacular gardens, wonderful art and furniture, and a private golf course. He also owned a yacht with a twenty-five-member crew that he kept moored on Lake Michigan.) Albert thought that Mary's offer was modern and endearing, but his "idea was that women should have independence and also be given everything in addition. They should do anything they want, earn money, and then you should give them everything as well, including anything, whatever it was that other women had." Not long into their marriage, Albert gave Mary a check for a million dollars. He felt she "shouldn't be without a large amount of cash." That would be the equivalent of $16 million in 2014 dollars—a large amount of cash for the sugar bowl, indeed.

There was plenty of money where that came from. A prospective client who hoped to secure Lasker's services would be required to cough up $1 million ($16 million in 2014 dollars) before Albert would consider taking on the new account. Clients with more substantial projects would have to put up $3–$4 million ($48–$64 million today) before he would get to work. Lasker was rich, and he was generous—unlike Mary's father who pinched pennies when there was no need to do so. (Mary described her father's parsimony as *infuriating*.) Though Mary had come into a fortune when she married Albert, she wasn't interested in buying things, per se. By then, Mary was far more interested in how to use money to achieve a goal than how to spend it on, say, another hat. Her low opinion of doctors had not softened in the least, and she felt she had a score to settle there. Both of her parents had died of stroke in old age. Nevertheless, Mary "was deeply resentful that nothing could be done to help" them at the end. "This I bitterly resented," she emphasized. It made her furious when doctors tried to reassure her that they could have done nothing more to prolong her mother's life, that her death was "an act of God." She vowed to work against "the will of God," and boasts in her oral history that *she succeeded in doing so*. As far as Mary was concerned, doctors were "small professional men," and what was needed was to "fight the lethargy of doctors as a group."

Mary had no intention of getting lost in the ranks of the idle rich; and besides, she had an ax to grind with the medical profession—it was not as good as it could be and didn't care to improve—and she intended to change that. During the first year of their marriage, she donated the equivalent of $40 million in 2014 dollars of Albert's fortune to a variety of medical philanthropies: arthritis, tuberculosis, public health, birth control, and cancer. Flooding the cisterns of her favorite charities with her husband's money produced a flood of grateful recipients who sang her praises and came back for more. Proclaiming herself an evangelist for medical research, she confessed in the passionate tones of the born again, "I am opposed to cancer and heart disease and stroke the way I am opposed to sin." She then distributed huge sums drawn against her husband's account as proof of her conversion.

Albert Lasker wasn't a stranger to philanthropy. He had tried it and didn't like it very much, being somewhat discouraged when his money did not produce tangible results. In 1928, he had donated $1 million to the University of Chicago to create an institute on aging called the Lasker Foundation. It suffered from a lack of leadership and limped along until the University of Chicago persuaded Lasker to release the funds for general use, which he did. He also donated $50,000 to the American Society for the Control of Cancer in the 1930s in memory of his brother, Harry, who had died of cancer. But Albert was a man who demanded product, not promises, in return for his philanthropy, and he soon tired of what he thought was the listlessness of charitable enterprise.

Mary had a hard time convincing him to take another look: "He knew absolutely nothing about health," but "he went along with me because he thought, you know, that it would be easier to go along." She argued that Albert's initial donation probably wasn't enough to produce tangible results. She was convinced that if a little money didn't yield the desired effect, more money probably would. Curious about where all his money was suddenly going and what it was being used for, Albert asked Mary what she hoped to achieve with her largesse. Her list was longer than his fortune was deep. He suggested to her shrewdly, "You don't need my kind of money, you need federal money, and I will show you how to get it."

Albert Lasker knew exactly how to get money out of Congress. He had been the head of the government shipping board in 1922 and again in 1933. "He knew a great deal about legislation and the mechanics of legislation and the psychology of politicians," Mary said. "He hated Washington," but "he knew what it took to get anything done there." More importantly, Albert knew that *only* Uncle Sam had the kind of money that Mary was looking for—a limitless supply, that is. Albert did not have to tell her twice. The chance to siphon off the US Treasury to set everything right in medicine was simply irresistible. Mary forever after justified her mining of the public coffers by saying, "Federal money is only our money in another pocket." And she trained her sights on getting as much of it as possible.

Albert's strategy for getting Congress to pay for Mary's philanthropy was rolled out in a series of seemingly innocuous steps that took place gingerly over several years. The first cunning step in the strategy to unleash his wife on Uncle Sam took place in 1942 when he created the Albert and Mary Lasker Foundation. He used some of his money to seed the philanthropy and augmented this with donations wrung from wealthy friends. His next step was nothing short of brilliant: he created three Lasker Awards, one each for medical research, medical leadership, and medical journalism. The vast domains that these three awards encompassed were then slowly turned to the Laskers' advantage. The Lasker Awards—leather-bound certificates large enough to be seen from the back of a room—were bestowed on individuals who were seen as the most prominent leaders in their field. It was a top-down seizure of the territory. The Lasker Awards were accompanied by handsome checks and were given once a year at grand ceremonies held in glittering venues. The press accommodated the Laskers' entry into philanthropy by covering their every move, and so everyone was bathed in a lovely light. It seemed like an aside, but it was part of the plan to persuade the recipients of the Lasker Awards that government should take the lead in funding medical research. That was an easy sell.

The next phase in Albert's strategy was to convince the public that federally funded medical research was in the public interest. That idea was easy to sell, too. In the wake of the New Deal, people increasingly began to look to the government to save the day. And increasingly, the government was there to bail them out. Albert had a ready audience, and he knew just whom to tap in the media (*New York Times*, *Reader's Digest*, Ann Landers), to carry his message. He argued that medical research was too big an undertaking to leave to individual universities and institutes. Problems as big as heart disease, stroke, diabetes, arthritis, and cancer required big, government-led solutions. There was nary a soul who disagreed.

Albert Lasker's connections in the media were deeply, tightly woven. His advertising agency paid millions of dollars to radio and television networks, as well as print media, every year to market a range of products from toothpaste to refrigerators. Cigarettes were an especially lucrative product line that provided a river of money for the networks year in and year out. Having saturated the market for men, Lasker dreamed up a clever way to market them to women: "Reach for a Lucky instead of a sweet," he said, to keep yourself thin and attractive. It worked. The networks were more than happy to do Lasker the favor of featuring whatever pet project Mary happened to be up to at the time. Was it cancer? Was it heart disease? Was it arthritis? No problem whatsoever. All Albert Lasker had to do was pick up the phone to set his minions chatting up whatever message he wanted poured into the public ear.

Radio, TV, and journals were soon singing the praises for a greater role of government in funding medical research. Having used the media to work the public up into a lather, Albert's next step was to leverage public opinion to wring appropriations out of Congress to fund the research. That was easy too. It was simply a matter of putting everyone on Capitol Hill on the Lasker payroll by donating *generously* to political campaigns. However, the money did not flow until *after* the quid pro quo promises had been delivered. Once congressmen learned that they were paid after the work was done, they worked hard for their Lasker money. They then returned to see how they might be of greater service in return for getting more. The favors given in return for campaign contributions amounted to two things: (1) increased funding for medical research and (2) supervisory councils, commissions, and committees where the Laskers and their friends could serve so that they could exert control over exactly how the funds were spent.

In summary, Albert Lasker's strategy involved spending a bit of his money to increase his visibility and access to power; using awards to create a tribe of the most important players in the field of medicine; and paying Congress to play his game, using campaign contributions to wring appropriations out of Congress to fund medical research and maneuvering allies into positions where the Laskerites could control how the funds were distributed to members of the tribe. The plan was pure genius, and it worked beautifully. Lasker boasted that he could take thousands of dollars of his money and turn it into millions of dollars of federal money. In truth, he turned it into billions. And that was the morning and the evening of the first day.

In 1942, when the Laskers started their private/public partnership, Congress appropriated approximately $2 million to the Public Health Service (PHS) and the National Institutes of Health (NIH) for their work. The majority of these funds were spent on basic science, sanitation, and the prevention of infectious disease—the major public health threats at the time. This kind of work is referred to as *primary prevention* because, well, it aims to prevent a disease from occurring. (*Secondary prevention* refers to preventing a disease that has already taken place from recurring.) Mary Lasker was not keen on primary prevention. She believed that patients died needlessly for want of better treatments, not want of better knowledge of the causes of diseases or how to prevent them. Believing that primary prevention, with its emphasis on basic science, was a waste of time, she came to the conclusion that treatments, not prevention, should be the nation's highest priority.

The kind of work that concerns itself with diagnosis and treatment of disease is referred to as *categorical research*. This involves the practical application of medical knowledge to clinical problems, like how to provide nutrition to a patient who cannot eat. The NIH and the PHS were doing categorical research, to be sure, but not enough to suit Lasker. Like her husband, she wanted to see more in the way of results. And like her husband, she had no way of understanding that knowledge drives results. (His expertise was advertising and manipulation; her expertise was art history and manipulation.) Mary Lasker insisted that the increased appropriations of federal money to support medical research be directed toward categorical research rather than basic science or primary prevention. Whenever she used the term *medical research*, what she meant was categorical research and nothing else. Lasker argued that the only thing standing in the way of real progress in medicine was the will to move forward, an accusation that's hard to rebut, especially when the accuser has everyone in Congress on her payroll. Mary Lasker was always annoyed when anyone disagreed with her: not everyone conceded to her husband's tendency to "go along because, you know, it's easier to do so." She locked horns with many men in the upper strata of government administration, especially the head of the Public Health Service, Thomas J. Parran Jr., MD: "I think the whole idea that you should do anything new about anything was a surprise to him and to everybody else" who worked with him. In demanding forward movement at whatever cost, Lasker insisted that the cart be put before the horse, that categorical research should go ahead of basic science. Having rearranged this fundamental principle of progress to suit her agenda, she cracked the whip and demanded speed. And Congress obliged in providing fuel: money in the form of appropriations. Naturally, the men involved in categorical research were delighted with this rising tide. Hospitals and drug companies made bigger boats to catch the passing waves. Patients were farther downstream, but they listened to the radio and watched TV and read *Reader's Digest*, and they watched for signs of change and hoped that they would be swept up and saved in glory too. "Any day now" became the refrain of every new song sung about treatment of disease. Of course, there *was* progress being made at the NIH, the PHS, and elsewhere in the country. But Lasker's latest ad campaign left the impression that because the government was taking the lead on categorical research, the rapture was expected soon.

Politicians took their checks and climbed aboard the Lasker train. They cheerfully, and then gravely, declared how concerned they were about the fight against disease. With public opinion filling their sails with consent to increase appropriations to the NIH and the PHS, Congressmen signed their names to one spending bill after another, as fast as the ink would flow. Once set in motion, this skillfully engineered mechanism that Lasker had designed created an upward spiral of self-reinforcing federal spending that grew larger by the year. It would have been unthinkable to pull back. An abomination to say, "Stop!" All were invited to gorge at the buffet table. And for the most part, those who "ate" became the Laskers' friends. That was the beginning and the end of the second day.

Mary Lasker personally cultivated every lawmaker in Washington, DC. She also kept an eye out for anyone who looked like they were headed there. In no time, Capitol Hill was tethered to her purse and answered to her call. No congressmen, senator, president—or his wife—went unsolicited. She had her favorites—Senator Claude Pepper of Florida was one of them. As I mentioned, she had two objectives: (1) persuade elected officials, by way of campaign contributions, to increase federal funding for medical research and (2) maneuver into positions where she and her "little lambs," as they were called, could direct how the money would be spent. One by one, Lasker collected political chips in return for cash, which she later exchanged for seats on committees and panels *that she created* so that she and her friends could direct how federal money for medical research was spent. In 1946, she told Albert that the American Cancer Society (ACS) should testify before Congress that $100 million was needed to fund cancer research. Albert had a hard time convincing the ACS of the wisdom of this extravagant request. He finally won them over when he told the members of the board that any increased appropriation that took place in Congress following their testimony could be claimed as an extension of their (good) work. That did the trick. Members of the ACS testified before Congress in hearings that *Mary Lasker arranged*, but the effort failed to secure the money they were hoping for. However, the precedent was set: the ACS began to see the federal government as an extension of itself, and others came to believe the same.

In his book *Cancer Crusade* (see Bibliography), the historian Richard A. Rettig has said of Mary Lasker, "Her presence was ubiquitous, her influence great. . . . She is, without question, a remarkable political figure" who "has a keen sense for the use of money in the pursuit of her objectives."[2] Lasker freely admitted the power of her purse. She describes in her oral history just how much money she spent currying the favor of politicians. Apparently, she had no qualms about flaunting campaign finance laws aimed at preventing such abuse. She simply ignored them and continued on her way. In one chapter of her oral history, she offered the following itemization of expenses:

> To the Vice President (Lyndon Johnson) and the President (John F. Kennedy) I contributed $50,000. You see you can give only one of them a check up to $3,000 made out to a state or national committee. But actually there's no legal way you can give them a large amount of money, except by making a check out to a variety of organized committees. . . . I gave a great deal of money to individual Senators and Congressmen through their committees. . . It's really an extremely awkward way to have to contribute. If anybody wants to give more than $3,000 he has to do it that way. It's ludicrously complicated because it can be gotten around so easily. Everyone accepts it. It's a way of getting around a limitation.

Lasker proudly admitted that she "was always interested in how to get something done by indirection." Her skill with indirection, greased with tons of cash, gave birth to a market-driven medical research economy, which she systematically organized around the concept of a cure—for heart disease, for arthritis, for stroke, and finally, for cancer.

The subtlest aspect of Lasker's strategy consisted of arranging for the creation of advisory committees, councils, and commissions. Congress ordered them at her request, and then it funded them and paid for their reports, which were used as testimony for an increase of funds for the next go-round. These advisory committees and so forth were composed of lay citizens (her friends) and scientific experts (her lambs). The agendas were set well in advance of the creation of the committees, which she packed and controlled. The advisory committees issued reports that argued in favor of a preset agenda, the need for more money for arthritis, and this is why. The advisory committee would present its report to Congress during a formal hearing. An integral part of the spectacle aimed at raising the emotional temperature of the crowd involved bringing in "citizen witnesses" that were trained to whip up hysteria in support of Lasker's agenda.

Citizen witnesses were handpicked delegates who were recruited and rehearsed for the task of appearing before Congress at hearings that Lasker arranged to make the case for spending more money on medical research. Citizen witnesses were selected from two bolts of cloth: (1) academicians who supported Lasker's view that the government should take charge of medical research and (2) powerful friends that could enter a room and, by their very presence, command it—in the way that, say, Bill Gates can sound like an expert when he talks about malaria. Citizen witnesses were "selected for their 'evangelistic pizzazz.' Put a tambourine in their hand and they go to work."[3] Citizen witnesses were not there to testify ad lib. They were provided with finely wrought scripts from which to draw their comments. The Lasker Foundation issued what Mary Lasker called the *Big Fact Book*, a compendium of data and statistics, charts, and graphs that supported the demand for more research. It was reissued every two or three years and was used as a reference volume to draft scripts for Congress and the general public. If the topic was cancer, for instance, the American Cancer Society, which did her bidding, wrote the scripts using the information in her *Big Fact Book*. Other diseases required other societies and other citizen witnesses, but the blueprint remained the same. Lasker knew just how effective it was to put three or four dressed up, tuned up, and wound up men in front of Congress, saying, "If all of them are successful doctors and scientists who agree . . . and call for big money . . . the impact is significant."

Richard Rettig describes Lasker's strategy deliciously in a chapter in *Cancer Crusade* entitled "The Benevolent Plotters":

> Several important elements in the Lasker style of operations can be identified. . . . First, Mrs. Lasker has substantial personal resources—money, time, commitment, the capacity to confer status on others—which she has used in a focused and skilled manner to further both general and specific ends. Her own resources have been augmented through the access to political leadership and the press and broadcast media that she has so assiduously cultivated. Second, Mrs. Lasker has frequently acted through an elite group of long-standing associates whose underlying commitment is to the support of medical research. . . . This elite has functioned oftentimes . . . through a blue-ribbon panel of committed experts. Though always including medical and scientific professionals, prominent lay representation has been a guiding organizational principle. . . . Finally, when mobilized, Mrs. Lasker and her friends have displayed great capacity to overwhelm the opposition through a focused, highly publicized, over-simplified, dramatic appeal to the public and through skillful tactical maneuver within the political process.[4]

Mary Lasker was recognized by all as a lobbyist extraordinaire. She was always in the background, but her hand was clearly visible nonetheless. She would have her citizen witnesses plead with Congress not to let them die. She would have them say that heart disease and cancer were completely avoidable. If only the congressman would vote in favor of spending more money, no one would have to die. Everyone could be saved. The money would "wipe away all tears." She was a terrific stage director whose repertoire was melodrama packed with facts; and it worked.

Lasker had learned from the success of the March of Dimes against polio how effective it was to focus the public's attention on one disease at a time. By sinking her claws into one disease and then another, Lasker slowly but surely recruited vertical segments of society to her side. If you had heart disease, she was on your side. If you had stroke, she was on your side. If you had cancer, she was on your side. Each group of patients was singled out and made to feel special, for a time. Each had its advisory committee, funded by Congress, its citizen witnesses, and its special appropriation of federal money. Lasker rode the circuit, stopping at each disease along the way, then circled back to accelerate the pace. Everyone was happier with every round. In no time, Mary Lasker was seen as a crusading Joan of Arc for medicine, per se.

In 1944, when the Laskers set out to flash their teeth at the US Treasury, the research budget for the PHS and the NIH, which included the National Cancer Institute, was approximately $2 million a year. Three years later, their budgets were up by a factor of five, to $10 million. The agencies were delighted to see more money pouring down Capitol Hill in their direction. The fact that the wind was blowing from Mary Lasker's direction didn't seem to bother them much at all. They were happy to see more money. However, the men who ran the PHS and the NIH were not inclined to take unsolicited advice about how to spend this money, especially when it was put forth in the most vehement terms by lay activists wearing furs and diamond brooches who sailed in on yachts and private planes. Certainly, these scientific leaders—long-time officers in the PHS—had no intention of taking direction from citizen witnesses regarding the setting of priorities for scientific research. These men knew exactly what they were doing, even if Mary Lasker thought they were all "small professional men." Lasker took the opposite view. She felt the men who ran the NIH and the PHS needed to be closely supervised so that the money they were given as a direct result of her lobbying was spent properly—on categorical research.

The Laskers began to set the talons on other philanthropies, as well, starting with the American Birth Control League. Mary Lasker had been a strong supporter of its founder, Margaret Sanger, and had donated small sums to the organization over several years. But after her marriage to Albert, she became more actively involved with the league, eventually becoming a member of the board of directors. Albert didn't care for the name of the organization, saying that the term *birth control* was offensive to the husband's authority over his wife and home. He felt the term implied that a husband ought to defer his sexual appetite in deference to his wife's preferences for family planning. Since the prevailing opinion at the time was that a husband had a right to the marital bed whenever it suited him, Albert thought that the American Birth Control League was bound to offend men. Albert, who had once donated $10,000 to Sanger's organization, believed that if the words *birth control* were dropped from the name of the organization it would attract more men, and therefore bring in more money. Mary Lasker agreed and suggested to the board that the name be changed in an effort to make the organization more broadly marketable to the public. Margaret Sanger was incensed. Birth control was her mission. It was the whole point of her campaign and her crusade. Sanger was vehemently opposed to any change in the name of the league that she had founded and had fought for, including several stints in jail.

Sanger might have saved her breath. Mary Lasker succeeded in getting the name of the American Birth Control League changed to Planned Parenthood. Having left her mark indelibly, Lasker resigned several years later. She said she was frustrated with the press. They kept pushing back, resisting a discussion of the rights of women to control the number of children they had. Most of the resistance came from the Catholic Church, the rest came from recalcitrant and stubborn men. Mary Lasker tried to iron out the crease she'd made in her relationship with Sanger by giving her a Lasker Award in 1945. Despite this overture, Sanger went to her grave resenting Mary Lasker for what she had done in erasing *birth control* and rewriting Sanger's legacy to fit a Madison Avenue man's preference for *planned parenthood.*

Next, the Laskers turned their attention to the American Society for the Control of Cancer (ASCC). The ASCC was founded in 1913 by a small group of physicians and surgeons who were interested in cancer. Their mission was to "disseminate knowledge concerning symptoms, treatment, and prevention of cancer; to investigate conditions under which cancer is found; and to compile statistics with regard thereto." In a nutshell, their mission was education of medical colleagues who were less familiar with the diagnosis and treatment of cancer and, more broadly, the lay public that knew next to nothing about the disease. The ASCC expanded its membership and its reach over the next fifteen years and, in 1929, it elected Clarence Little as its director. Little was a devoted, charismatic, and enthusiastic leader of the organization. He was still at the helm eight years later when, in 1937, he launched a major initiative to educate women about cancer. This was really quite visionary for the time.

To this end, Little organized the Women's Field Army, a corps of two million volunteers from thirty-nine states whose job was to raise $2 million to fund a broad educational program to teach doctors and women how to look for early signs of cancer and, if needed, direct patients to cancer experts where they could receive the very best treatment available. Little made the cover of *Time* magazine on March 22, 1937, at the launch of his campaign. He is grinning in the photo in the manner of Clark Gable in *Gone with the Wind* while he lights his pipe, a professor with a twinkle in his eye. The accompanying article, which was glowing in every respect, described the Women's Field Army as "the largest evangelistic movement ever loosed against a disease."[5]

Little recruited two very prominent women to his side to help him spearhead the campaign: Mrs. Grace Morrison Poole, chairwoman of the General Federation of Women's Clubs, and Mrs. Marjorie B. Illig, head of the federation's health division. (Illig was a substantial woman with serious credibility on her resume: she was a doctor and a radiologist, and she was married to an executive of General Motors. When she spoke, everyone listened.) Little's objective in his massive national campaign on behalf of the ASCC was to educate all forty-five million women in the United States about the early signs of cancer and advise them where to go for care should the need arise. He intended to carry out the campaign through a series of lectures that would be sponsored by the member organizations of the General Federation of Women's Clubs. His goal, emblazoned on the cover of *Time* beneath his portrait, was to get "all the women talking about cancer" through "mass meetings, lectures, radio broadcasts, newspaper and magazine articles: and the distribution of tons of literature."[6]

When asked why he was so passionate about his new project for the ASCC, Little explained:

> Why do I feel so deeply about it? Because I have both experienced, understood and, I am afraid, caused too much suffering, and hate it. Because my own father died as a result of cancer. Because perhaps whatever ancestral desire I have to explore the unknown is appealed to by the research work and the wish to be a "crusader," which almost all of us have, if given a chance to express itself. Finally, because I believe that Americans will be happier and saner if they combine in fighting a scourge like cancer than they will be if they continue to fight each other for money and power.[7]

Ah, yes, the perpetual fight "for money and power." It was headed right for him. Mary Lasker showed up in Little's office one day in 1943 demanding to know how much money the ASCC was spending on cancer research. Little replied that research, per se, was not part of the society's mission. He explained to Lasker that its focus was education of physicians and the public. He had certainly fulfilled that mission, and then some, with the success of the Women's Field Army—*women were talking about cancer*. Little explained to Lasker that cancer research was more effectively carried out at academic institutions where laboratories, scientists, and the equipment necessary for this kind of work were already in place and thriving.

Lasker was appalled. She felt that the ASCC ought to get involved in medical research, that education was simply not enough. She left Little's office, vowing to change the ASCC's mission and its course. But she kept her huff to herself for the time being. She contributed $5,000 to the ASCC so that Little could print up another batch of cancer education pamphlets. And then, in a shrewder move, she made a donation to the ASCC the following year, 1945, with the stipulation that the money be used to establish a formal research fund. The fuse was planted. It remained to be lit. As Lasker later admitted, she "wanted to put dynamite in the organization." She also persuaded Albert to leverage a deal he was cutting that involved the sale of his stock in the interests of changing control of a company in the midst of a takeover so that he would sell his shares provided the company donated $50,000 a year for five years to the ASCC *specifically for cancer research*. This was no skin off Lasker's back, and it secured more money for the ASCC—quite a lot, in fact—that helped to advance his wife's designs on the society.

Little was delighted with the contributions made by the Laskers, and he called Mary and asked her if she would speak to Albert to see if he would be willing to serve on the board of the ASCC. "I thought my husband would not be interested in doing this as he did not like to work with organizations. . . . God knows he couldn't stand such nonsense." Albert suggested his friend and protégé at the ad agency, Emerson Foote.

Both of his parents had died of cancer, and he was only too willing to serve. Mary Lasker convinced Foote that the ASCC needed a big fundraising campaign, and he agreed to get one going. She also told him that the name of the organization should be changed. And it was, for there was no one who really cared one way or another as long as the message was clear. And so, the American Society for the Control of Cancer, arguably a mouthful, became the American Cancer Society (ACS)—a title short and sweet.

Foote began working on the members of the board of the ACS to initiate a huge fundraising campaign. Mary Lasker, who said she was "always best if I'm behind the scenes," offered to pay the salary of a fundraising manager she knew who had been involved in political campaigns in the past, provided the ACS agreed to give 25 percent of the money raised to cancer research. No one had a problem with that, and it was agreed upon. Lasker brought in Leo Casey and paid him $18,000 a year—a fortune in 1945. He didn't seem to work out, was fired, and was replaced by Eric Johnston, who made it rain like never before. Eric Johnston was head of both the United States Chamber of Congress and head of the Motion Picture Association of America. He knew as many people as Albert Lasker, if not more.

Like Lady Macbeth, Mary Lasker pitched in to help. She called her friend, Mrs. James Monahan, medical reporter and editor of *Reader's Digest*, who wrote under the pseudonym Lois Mattox Miller.[8] She asked that a series of supportive articles be written explaining the untapped potential of cancer research. Mrs. Monahan obliged, even running a direct appeal for donations to the ACS at the beginning of this blitz. In October 1944, *Reader's Digest* featured an article that had been published in *Hygeia* (written by Bernadine Bailey), "Today's Cure for Cancer."[9] It focused on female cancer, the leading cause of death in women at the time. (Today, it is heart disease.) "If cancer is detected in the beginning stages, 100 percent cure is theoretically attainable," Bailey writes. "Actually, it is delay, ignorance, and fear that cause most of the deaths from cancer today. When these are overcome, the war against this once-dread disease will be largely won. In the final analysis, it is up to every individual. You—and you alone—can beat cancer."

It's worth mentioning that in her article Bailey was emphatic, using italics to help make her case, that cancer "is *not contagious or infectious.*"[10] She was obviously unaware that Bittner's mouse breast cancer virus had, by then, been seen with the electron microscope, which *confirmed* (italics mine) that cancer—at least in mice—was an infectious disease.

The following month (November 1944), staff writer Paul de Kruif, PhD (a close friend of Lasker's), planted another cancer story in *Reader's Digest*, entitled, "Fifty Thousand Could Live." It begins, "Excepting pneumonia, there is little question that cancer is now the most readily curable of our major diseases."[11] Reading from Lasker's catechism, de Kruif said that early detection, surgery, and radiation therapy were gateways to a cure. Quoting "the grand old man of American medicine," pathologist Ludvig Hektoen, MD, de Kruif continued his epistle: "Dr. Hektoen believes that science may someday find a simple chemical cure for cancer . . . but 'you can't satisfy the present cancer victim by proving to him that research into the future may bring a cure. You've got to treat him with the means at hand, and those means are really magic.'"[12]

Thanks to articles like these, the notion that doctors were suddenly pulling rabbits out of the hats and curing cancer patients left and right took root across the nation. It didn't take much watering to turn this notion into a weed of expectation. Through clever marketing and strategic messaging, the hype and hope for an imminent cancer cure that was promulgated by the *Reader's Digest* articles led to a sharp rise in donations to the American Cancer Society. Eric Johnston called on his friends in the motion picture industry and arranged for direct appeals to audiences sitting in the movie theaters. Albert Lasker called his media friends and got everyone on radio and TV talking about cancer. Money started to roll right in. In 1943, the Women's Field Army had raised a total of $356,270 for the American Society for the Control of Cancer. In 1945, the year following the Lasker-orchestrated extravaganza, the society took in $4,292,000—an increase of 1,200 percent. To put that into context, imagine if your weekly salary was $356 one year, and the following year it was raised to $4,292. The difference would be palpable, and it would make you very happy, indeed. As stipulated by Lasker behind the scenes, 25 percent of the $4.2 million raised by the American Cancer Society that year was siphoned off for cancer research: $1,073,000! No one complained.

Although the agreement to use 25 percent of the funds raised by the ACS in 1945 did not have a "lean and hungry" look, there were knives buried in the folds waiting for their moment to appear. The moment came as soon as money started coming in. Lasker had convinced Albert to join the board, even though Foote complained that it was "a constant effort to keep the doctors in line." Mary Lasker wanted to change the composition of the board of directors at the ACS: she wanted half the members to be lay citizens, by which she meant her powerful friends. Foote, Johnston, and Albert agreed to see what they could do. Foote invited Mary Lasker to attend the board meeting where he brought the subject up for discussion and a vote. "The doctors were horrified," she recalled. "They really didn't understand what happened to them." Historically, the members of the board of directors had been medical experts—physicians, surgeons, and academics of one stripe or another. The Laskers' argument for making such a drastic change in the composition of the board went something like this: "We brought in all this money, and we can bring in a whole lot more. We are experts in managing the public and its money, and we ought to have a greater say in how this organization is run." But they didn't really want a greater say; they wanted to control the whole thing: a coup never ends in an equal distribution of power.

The board members were stunned. Mary Lasker recalled that Little, especially, "was very difficult." The board members couldn't understand how lay citizens—who were not "lay" at all but were wolves dressed up like crusaders—had come to believe that they could make rational, informed, scientific decisions about something as complex as cancer research. Arrogance is seldom restrained by ignorance, and it certainly was not in this case. As Richard A. Rettig describes in *Cancer Crusade*, Mary Lasker was convinced "that progress in the war against cancer was too important to be left in the hands of conservative physician-scientists." As Rettig reports, and history confirms, "a bitter fight ensued."[13]

The physicians who sat on the board of the American Cancer Society may not have been as familiar with Shakespeare's *Richard II* as they were with, say, Sir William Osler's *The Principles and Practice of Medicine*, but they certainly knew an insurrection when they saw one. Albert Lasker, ever the smartest guy in the room, recognized the fierce and mounting opposition that risked dissolving into chaos, and so he moved to the ropes to take the blows, acting as if he and his team were defeated. But he was really only buying time, for he arranged for another friend to join the organization, Elmer Bobst of Hoffman-LaRoche. At that point all hell broke loose. As Bobst later recalled, "I decided that the first priority was to move aside the scientists and physicians who were in administrative control of the organization. They were good men, but they were not experienced leaders, and they were not getting results. I wanted majority control to be in the hands of qualified lay leaders. The physician members could form a scientific committee to make recommendations about scientific matters and advise the executive committee."[14]

Mary Lasker describes the doctors' response as "incredibly obstructionist; incredibly mean and difficult." She thought she understood their motive in resisting her proposed change in the composition of the board: "They couldn't stand men with ability and business know-how." The Foote and Johnston and Bobst and Lasker crowd used their influence with other members to move the beleaguered Little around the executive board until they had the king in checkmate. Mary Lasker said that it was "with great fighting that Little was persuaded to resign." Actually, the game was up when Little was told that if he didn't stage an honorable "suicide" and resign, they'd have his head instead.

Little left the American Cancer Society quietly and gracefully, although reports of the battle that led to his demise were by then broadly known. Although his leaving was not a mortal wound, it was humiliating for Little to be set on the curb like so much trash. When asked why Little was asked to leave, Mary Lasker said it was because he "couldn't get along with anybody and was totally obstructionist." In fact, she believed that "none of the men there knew how to run the organization."

To review, in the years Little served as director of the American Society for the Control of Cancer, he had done an impressive job raising awareness about cancer and providing solid, useful information for the medical profession and the public, especially women. Little was generally seen as a brilliant researcher, a man highly regarded by his colleagues. He had risen precociously through academia to become the president of two universities before he entered middle age. He had fulfilled his dream of heading up his own research laboratory, JAX Lab in Bar Harbor, Maine. He was director of the ASCC for many years, the most important cancer society in country, where he spearheaded the national campaign to get women talking about cancer. As a result of his development of inbred mice, Little made a major impact on the course of cancer research, a contribution that has only become more important over the past hundred years. Sadly, he was sent packing on trumped-up charges, judged incompetent not by a jury of his peers but by a group of "Mad men" and a dressmaking pattern queen, who were convinced that they knew better how to run the show.

Over the years, Mary Lasker moved up the ranks of the American Cancer Society, variously serving as its director, chairman, and trustee. Eventually, the entire operation was virtually at her disposal. In the late

1940s, control of the society was wrenched from the physicians who created it and put into the hands of private businessmen who thought they knew better how to run organizations. Albert Lasker resigned from the board saying he was "frustrated with the doctors and the organization." One wonders why he and his wife didn't just start a society of their own and then leave the doctors at the American Society for the Control of Cancer alone to fulfill their mission as they saw fit.

Being a woman of unfathomable energy and drive, Mary Lasker simultaneously directed her attention to the National Cancer Institute (NCI). It put up a much bigger fight. Its director, Kenneth M. Endicott, MD, had no intention of taking orders from a rabid activist, and he had the entire Public Health Service—a division of the armed forces—on his side. When Lasker demanded that the NCI tailor its research agenda to focus more of its efforts on categorical research, Endicott growled and came out of his hall like Beowulf to face her down. This sent her scurrying for cover; but of course, she called for her cavalry and suited up for more. Endicott was prepared for her on every front. She had a mighty hard time scaling the walls he built to defend the NCI. In an accusation that surely smacks of melodramatic hysteria, Lasker condemned Endicott's intransigence by declaring, "This man is a power drunkard unconcerned about human beings." The charge was hyperbolic and ridiculous. He was a better fighter, simple as that. Endicott, a proven leader and long-time veteran of the PHS, who had risen admirably in its ranks, simply refused to be tilted off his horse by the raving Mary Lasker.

On occasion, Mary Lasker's harsh judgments converged to the point of cruel. For example, she leveled her ire at a Mr. Fairby, a member of the Bureau of the Budget, the agency whose job it is to keep track of government money and make sure the federal checkbook is balanced at all times. Apparently, Fairby didn't care for Mary Lasker and didn't agree with her demands that the government spend more money on medical research year after year. Fairby wasn't always sure that it was a good idea for the federal government to take the lead on this, let alone pick up the tab for all of it. (The country had barely survived the Great Depression, had lost a fortune in World War II, and was struggling to recover.) These arguments not only didn't persuade Mary Lasker, they got in her way. She was convinced that Fairby was to blame when Congress refused to cough up all the money she requested. Fairby, she said, "was at the bottom of our difficulties. He was small-minded and anti-everything, probably because he had a shrunken arm and nobody was able to help him."

Even among the very rich, nightmares have a way of coming true. In June 1950, Albert Lasker was diagnosed with gastrointestinal cancer. He underwent surgery at the Harkness Pavilion of Presbyterian Hospital in New York City on July 5. Unfortunately, his tumor was aggressive. By the time the cancer was discovered, it had spread to his lymph nodes. Not long after, it spread throughout his body. He died two years later, on May 30, 1952. Mary was frantic throughout the ordeal. Desperate for help, hoping for a miracle, she cringed helplessly in anguish amid the impending doom. As in every case, it was an agony for Mary to see her handsome, enchanting husband, the icon of her dreams, reduced by inches and then degree to the pulp of a living corpse. In the end, death was the only remedy. The last two years they spent together dug a deeper trench into Mary's furrowed brow. She had suffered horribly with illness as a child and had grown up resenting doctors and distrusting medicine. To witness a loved one, a family member, a friend, or a cherished partner brutally corrupted by disease, eaten slowly from the inside out by spreading, gnawing cancer, is a wrenching torment from which there is no relief. For Mary Lasker, who framed disease in terms of sin, her husband's suffering was particularly unbearable. Everyone had failed her. They had failed her husband. Everyone had sinned. Everyone was thus condemned.

Years later, the wounds were still visible as scars. In the photo that accompanies her oral history, which can be found on the Columbia University website, one sees a stiffly coifed, attractive, chic, well-tailored woman in middle age. Lasker looks pleasantly into the camera and appears to lean forward just a bit, as if reaching for the lens to make a point. But look more closely and you will see incongruence in her face. Her smile, which is less than convincing on further inspection, is parked beneath a frown. The image conjures up a frantic child who has been left alone in her crib to cry for far too long. When rescue finally comes, the child is both relieved and furious. Resentment boils underneath. Thus, the smile and the frown and the pitch forward to be rescued. This photo captures her courageous attempt to assuage a protracted grief that was made worse with Albert's passing and remained ever more energized by anger than by compassion for herself and others.

Chapter 10

Garb

An object in motion is more easily deflected than one at rest. The faster it is moving, the more deflected it will be when disturbed from its course. As an example, imagine a baseball launched from the arm of Mariano Rivera, the Yankees' former closing pitcher. The ball rockets toward home plate at 92 miles per hour, his average pitching speed. If all goes well, from the Yankees' point of view, the batter misses the ball, and the ball hits the mitt with the force of a truck. (I'm a Yankees fan, so we'll call that a strike.) The batter, if he hopes to hit this thing, has to start his swing before Rivera releases the ball from his fist. Only the very best hitters can pull this off and only about a third of the time.

Now, imagine that the batter gets the timing right, but only the tip of the bat strikes the ball as it whistles across the plate. The ball is not struck, but deflected; it ricochets high into the stands behind the batter like a particle from the Large Hadron Collider. Fans hear the crack, see the soaring mistake, and scramble and tumble to capture the stray. A frantic skirmish ensues as hot dogs, beer, and caps go flying. Rear ends and the soles of shoes come into view. This is part of the sport and part of the fun, but neither the pitch nor the swing does a thing to advance the game. It's not a strike or a ball or an out or a hit. It is nothing; you must try again.

This fly-ball scenario comes close to describing what happened in 1968 when Solomon Garb, MD, threw his screwball pitch of a book,

Cure for Cancer: A National Goal (see Bibliography), at the world. Batter, batter, batter, Mary Lasker, took a swing with her slugger and sent the book sailing high into the stands as the War on Cancer.

Dr. Solomon Garb was born in 1920, the year Mary Lasker made the switch to Radcliffe College. Garb received a baccalaureate degree from Cornell University in 1940 and earned his medical degree there three years later. He stayed at Cornell to train as a general practitioner and along the way developed an interest in pharmacology. After completing his residency in internal medicine, he remained on staff at the medical school for the next ten years. In addition to caring for patients and teaching students, Garb conducted animal experiments with a variety of drugs. He also experimented with cancer antigens (in other words, proteins made by cancer cells) that are sometimes unique to cancer cells, but are more often just normal proteins that cancer cells make in large quantities. Like all antigens, cancer antigens are perched on the surface of cancer cells like antennae fixed to the roofs of homes. Instead of picking up, say, radio waves, cancer antigens pick up chemicals passing by in the bloodstream (for example, hormones like estrogen). These chemicals then produce signals inside the cell that promote cell growth.

In 1962, Garb published a single paper about cancer antigens that are found on the surface of human leukemia cells. He discovered that when these cancer antigens were isolated (in other words, identified and extracted) and then injected into rabbits, the antigens lost their power to capture chemical signals and promote cell growth. Somehow, the cancer antigens had become inactivated. Garb discovered that once the rabbit's immune system came into contact with human leukemia antigens, the cancer cells could no longer grow. But why was their growth inhibited? Garb discovered that the rabbit's immune system recognized the human leukemia antigens as foreign and had made antibodies to them. These antibodies then neutralized the human leukemia antigens, making them ineffective (that is, incapable of picking up and reacting to chemical signals that would help them grow). Because rabbits were capable of making antibodies to human leukemia antigens, they did not develop leukemia. The rabbits' ability to make neutralizing antibodies to human leukemia antigens protected them from leukemia, something that, apparently, the human immune system was not able to do. However, despite the interesting nature of these findings, there is no indication that Garb pursued cancer antigen research further.

Later that year, Garb published another paper, in which he reported that rabbits were able to make antibodies to every protein that was found in human blood. Once again, Garb published no additional papers on this subject, so nothing more seems to have come of this line of inquiry either. Four years later (1966), Garb published another paper about neutralizing antibodies, this time in mice with breast cancer. Garb used JAX Lab's inbred mice (descendants of Lathrop's pet inventory) for his experiment. He reported that rabbits were able to make neutralizing antibodies against the breast cancer virus. Interestingly, this was a finding that scientists at the National Cancer Institute had reported, too. Again, Garb did not follow up this experiment with any further work on the subject. It's not known why his research on cancer antigens came to a halt. Perhaps he was no longer interested; or perhaps he ran out of money and could not find additional funding to support his work.

Judging from his short bibliography, Garb's primary interest seems to have been in the field of pharmacology. While on staff at Cornell Medical College, he experimented with a variety of drugs, particularly cardiac medications and tranquilizers like phenobarbital. From time to time, he received grant support—small sums of money—from the National Institutes of Health, but by the time he published *Cure for Cancer*, his federal funding had run out. It was never renewed.

There is no evidence that Garb developed any expertise in treating cancer patients, though as a general practitioner at a university medical school, he must have cared for some. Nor is there any indication that

Garb participated in trials involving cancer drugs or other cancer treatments. There is no record that he supervised or managed cancer research programs in animals or humans, or that he obtained any advanced degrees in public health, pharmacology, or oncology. Indeed, Garb's published record reflects an experimental dabbler, a curious physician with an interest in cardiac drugs and cancer antigens, who never quite found a subject that held his attention long enough to acquire breadth or depth.

Garb may have had a thin offering of papers published in peer-reviewed journals, but he loved to write nonetheless. He issued a series of monographs on a range of subjects, primarily written for nurses, medical students, and the like. Typical of his repertoire were a manual of laboratory tests and a glossary of common medical terms and abbreviations. He wrote a guide for disaster management. He wrote about the importance of nutrition in clinical medicine. His writing, like his research, cast a wide, albeit gossamer, net. In general, Garb consolidated information but didn't add to it. After a decade at Cornell Medical School, Garb moved to the University of Missouri, where he was made a professor of pharmacology. This was an interesting appointment given that Garb held no advanced degree in pharmacology and could only point to a handful of published articles in the field.

In 1968, Garb released his magnum opus. Using the space program's success as the inspiration for his book, Garb argued that if we could put a man on the moon in less than a decade, we could cure cancer within ten years too. He evangelized like a prophet, saying, "If the American people can accept the placing of a man on the moon and similar projects as important national goals, surely finding a cure or control for cancer is a reasonable and worthwhile national goal." Garb's book, launched on the eve of Apollo 11, resonated with an American psyche fully loaded and fully fueled for any farfetched plan. As Karen Blixen (played by Meryl Streep) whispered to Denys Finch Hatton (played by Robert Redford) in the movie *Out of Africa*, when after a day exploring the Serengeti together, he followed her into her tent after dinner and offered to help her undress: "If you tell me anything now, I will believe it."[1]

Given the enormous progress that was being made with drugs like penicillin, it's understandable that pharmacologists like Garb would be overly optimistic that a drug would soon be found to cure cancer. In the 1950s, Sidney Farber, MD, of Harvard University had pioneered the use of chemotherapy in treating children with leukemia, and he had achieved remarkable success. Many children who would have died in weeks lived for years; some lived into adulthood. A few were cured. Naturally, it was reasonable to assume that equally effective drugs would be found for other tumors in a matter of time. Unfortunately, researchers did not appreciate the important differences between cancers of the blood, such as leukemia and lymphoma, and solid tumors, such as breast, lung, and colon cancer. Their growth is different in ways that make them distinct from one another. Nor did scientists foresee the day when bacteria would become resistant to antibiotics, or a time when cancers would routinely become resistant to chemotherapy.

In *Cure for Cancer*, Garb presented a paint-by-numbers blueprint for curing cancer. He spoke as if he were an authority in the field instead of the professional spectator that he was. Garb's 303-page prospectus included a three-page budget with numbers rounded off to the nearest million. At the time (1968), the National Cancer Institute's annual appropriation was $185 million. Garb suggested that the government commit to curing cancer within ten years, which he estimated would require the federal government to increase this budget by 3,500 percent to reach the goal—$650 million per year to find the cure. Garb argued that because cancer was costing the country about $12 billion per year, his proposal for this drastic increase in federal funding was more than justified. Economically, it made sense. Morally, it was an obligation. Garb outlined a list of what were, in his opinion, the most promising "Cancer Research Projects." At the end of every discussion he provided a long list of references. There was nothing on this list that wasn't already being investigated elsewhere in the country. Indeed, the only thing new in his proposal was inflation for the cure.

Though Garb's book did not appear in the top two hundred books released in 1968, its greatest success was when it landed in Mary Lasker's lap. Her husband had died of cancer twenty years earlier, but it might as well have been yesterday. Mary recalled a time when, as a small child, she was taken by her mother to visit the family laundress who was recuperating from breast cancer. Mary was stunned when she was told that the woman's breasts had been removed. (One wonders at the wisdom of sharing such a gory detail with such a sensitive child.) Mary said, "I'll never forget my anger at hearing about this disease that caused such suffering and mutilation . . . I was absolutely furious."[2] Later, when she was traveling in Europe after college, her friend's mother was diagnosed with breast cancer. She died two years later. "This puzzled and infuriated me, because I found that so little had been done at the time about cancer."[3] Of course, this was patently untrue: thousands of scientists and millions of dollars, public and private, had been spent on cancer research by the time Lasker came on the scene. But she was convinced that support for cancer research was an entirely new idea, hers.

If Garb ever dreamed of being a one-man catalyst, his prayers were answered the day Mary Lasker read *Cure for Cancer*. He had her at "hello" when he said in the preface of his book that a cure for cancer could be found "within a few years." Lasker was enthralled with the notion of landing a cure on cancer in less than a decade. She immediately got in touch with Garb, and within a few weeks he was sitting in her home on Beekman Place in New York City. The two had several compass points in common: opinions that once formed never faltered, a sense of their own destiny, the conviction that others should get in line behind them, the belief that they were crusaders with a moral right to impose their views on the rest of society. As to the question of curing cancer in the time it took to go to the moon, Garb recalled that they agreed "on practically everything." Lasker vetted Garb's book with her good friend Dr. Sidney Farber, who concurred that a cure for cancer could be found in no time if the government were willing to get behind the mission as robustly as it had gotten behind the space program. Farber, the only one with any clinical experience on cancer wards, might have taken into consideration how important basic science was to any progress made in medicine, but he did not. Instead, Farber maintained that it was "not necessary, in order to make great progress in the cure for cancer, for us to have the full solution of the problems of basic research." (This is rather like planning a trip to the moon without first knowing the laws of motion.) Lasker took Farber's benediction of Garb's book as a blessing to move forward at warp speed. The next step was to launch the plan into outer space.

The secret sauce in Lasker's prize-winning recipe for yoking public money to private philanthropy was to create a citizens' committee—a handpicked group of lambs whose job was to cast the yoke in a favorable light. Once assembled, the committee would issue its report—always favorable to her design—complete with data from her *Big Fact Book* and supported by information provided by obliging organizations like the American Cancer Society. The citizens' committee's official report was then used to convince Congress, whose members were all on her payroll, that there was ample justification to fund the appropriation. Congress would then create its own committee to address the proposed appropriation, use the citizens' committee report to flesh out its own deliberations, and then report that it, too, felt there was ample reason to fund the appropriation. Once this was baked to perfection under the Capitol dome, an appropriation would emerge that Lasker and her lambs would maneuver to control directly and indirectly. It's worth mentioning that the downstream recipients of this largesse were pleased and increasingly accommodating to Lasker's demands. Indeed, most swerved to get out of her way, for they knew she'd return with loads of money.

Having decided that the time had come for the country to cure cancer, Lasker assembled the Citizens Committee for the Conquest of Cancer. The mailing address for the Citizens Committee for the Conquest of Cancer was the same as that of the Albert and Mary Lasker Foundation at United Nations Plaza in New York City. Emerson Foote, the man who spearheaded the ACS's hugely successful fundraising campaign in 1945, and Solomon Garb cochaired this private enclave.

Mary Lasker issued the committee's report in 1969. In it she alleged that "no large mission or goal-directed effort" had ever been made against cancer, and that finding a cure was reasonable, feasible, and necessary—"at whatever the cost." The report called for the establishment of a presidential commission for the conquest of cancer. Then, to be helpful, she included a detailed list of the people she thought ought to serve on the presidential commission. No surprise, Farber and Garb were on the list. The business community was well represented on Lasker's list too: the presidents of Itek, Arthur D. Little, United Auto Workers, ABC, Ford Foundation, and Bell Labs; a vice president of IBM; and a contributing editor of the *New York Times*. It's not clear why these businessmen who knew nothing about cancer research other than how to market it might be able to render an opinion about the merits of a ten-year race for a cure. Medical experts from the most prestigious institutions made her list too: M. D. Anderson, Stanford University School of Medicine, the Red Cross, Roswell Park Cancer Institute, Memorial Hospital, Columbia University, the Mayo Clinic, the University of Nebraska, and the American Public Health Association. Interestingly, Dr. Wendell Stanley of the Virus Laboratory at the University of California, Berkeley, also made the list.

The report's closing argument made a passionate case for landing a cure on cancer: "Diseases which were considered the 'will of God' have been conquered in our time. Tuberculosis deaths have declined nearly 90 percent since 1946 due to drugs, and polio is almost non-existent due to vaccines."[3] Finally, as if ten years was too long to wait for the rapture that Garb had prophesized, Lasker declared that a cure should be found in *five years*—by 1976, in time for the nation's bicentennial.

Though Lasker had made a pet of Congress, she realized that she would have to bring the public into the ring in a very substantial way if she hoped to get a billion-dollar appropriation for a cure passed in both houses and signed into law. The message that a cure was possible if only Congress would fund it would have to be planted, watered, and fertilized regularly in the public arena before she made her move on Capitol Hill. The idea had to become a self-evident fact, gilded with moral imperative, if she hoped to harvest an appropriation the following year.

No problem. She got in touch with her good friend Senator Ralph Yarborough from Texas. He agreed to appoint an advisory commission to the Appropriations Committee, which he chaired, in the Senate. She advised him, "There might have to be a special Cancer Conquest Administration, like the NASA administration, outside of the National Cancer Institute to get this done."[5] This was the first indication that she wanted to create a new cancer agency, apart from the NCI, answerable to her. Yarborough was in a tough reelection bid at the time and didn't give much thought to Lasker's request except to say yes and ask her for help with his campaign expenses, which she provided with a smile. Yarborough was in the fight of his life, up against a newcomer oilman, George H. W. Bush.

In October 1969, she was invited to have lunch with her old friend and former ranking Republican for Appropriations for the Department of Health, Education, and Welfare, Representative Mel Laird. "I thought I would talk to him about the idea of making cancer a national goal,"[6] she said. She buttonholed him and got him to agree to take the idea to President Nixon. On December 30, 1969, she received a letter from Laird saying that Nixon was interested.

Lasker then circled back to Yarborough. She told him that Nixon might be interested in setting up a presidential commission to consider the conquest of cancer, and what did he think of that? She cleverly reasoned that "if the President could be persuaded to appoint a Cancer Commission it would be easier because if the Commission came forth with some substantial recommendations and it was his commission, he would be more or less embarrassed about not following the recommendations." Yarborough made it clear to Lasker that he would take care of paving the way for a national commitment to curing cancer within ten years. There was absolutely no need to involve the president, as Yarborough had it well in hand. Lasker now had these two powerful men pitted against one another, which played well to her hand.

In November 1969, Lasker met with her old friend Elmer Bobst, who was a very close friend of Nixon, too. She asked Bobst to speak to Nixon, specifically to let him know that she would be asking for somewhere in the neighborhood of a half billion dollars a year to find the cure for cancer. Lasker went back to Yarborough and told him that he needed to get something organized to present to the Senate before Christmas. "Maybe this helped to make Mr. Nixon interested," she said. "Now I've got them both interested," she boasted.

The fireworks got underway on December 9, 1969, five months after the *Eagle* had landed on the moon. Lasker took out full-page ads in the *New York Times*, the *New York Daily News*, and the *Washington Post* to announce a "moon shot for cancer." She paid an advertising agency nearly $50,000 to create and run the ads—more than my parents paid for their solid-brick, center-hall, two-story colonial on an acre of land outside of Washington, DC. In a calculated ambush that Albert Lasker would have surely admired had he seen it from beyond the grave, the ad took the form of an open letter to President Nixon. Because she always preferred to get things "done by indirection," the letter was signed: "Citizens Committee for the Conquest of Cancer." In a shocking departure from protocol and respect, the letter was addressed to "Mr. Nixon." ("Mr. President" is the proper form of address. "President Nixon" would have been an acceptable alternative.)[7] Insult was followed by a shocking declaration: "YOU CAN CURE CANCER."

Good thing this was published in the *New York Times*, the *Daily News*, and the *Washington Post*, for it must have come as news to Nixon, not to mention everyone else in the country. The letter then thrust its sword to the hilt: "You have it in your power to begin to end this curse. . . . There is no doubt in the minds of our top cancer researchers that the final answer to cancer can be found . . . Dr. Sidney Farber, past President of the American Cancer Society, believes: 'We are so close to a cure for cancer. We lack only the will and the kind of money and comprehensive planning that went into putting a man on the moon.' Why don't we try to conquer cancer by America's 200th birthday?" the letter asked. The question lay on the page like an indictment that demanded a verdict, and then settled into the folds of the American conscience.

Lasker had nailed Garb's thesis to the White House door and tossed it on the stoops of millions of homes around the country. As planned, the public immediately got itself worked up into a lather. President Nixon, who had been charged with curing cancer in five years *if he wanted to*, raised his hands as if to say, "Don't shoot." But that was just a ruse, like his Checkers speech. He was every bit as shrewd as Lasker and more cunning. He was even more aggressive. And he was president. Nixon let out a grand hurrah and championed the idea as if it were his own. He called it *his* War on Cancer, a welcome substitute for the one that was incinerating his reputation in Vietnam. Then he did everything he could to get the country to view *him* as the leader of the cure. (Senator Edward Kennedy, a preeminent health care advocate, thought a race for a cure was a swell idea, too. He also wanted to be seen on the vanguard of the War on Cancer. In the hectic months that followed, Nixon and Kennedy sparred like rutting bucks in the shade of Lasker's strategy, while Yarborough rushed to get a hearing arranged in the Senate.)

In Nixon's inaugural address in 1971, he announced the War on Cancer and set the goal of finding a cure within five years, by 1976, in time for the nation's bicentennial. It was, word for word, taken from Mary Lasker's ad. In a comedy that Shakespeare might have written if only he had lived long enough to turn the farce into a play, Lasker had seized the idea of a race for a cure from Garb, only to see Nixon steal it from her when she threw it in his face. She didn't mind a bit: this is what political indirection is all about.

The promise of a cure may have ravished the general public, but scientists in the field—apart from a few like Farber—were frankly wary. In fact, when Farber and Garb tried to recruit other important signatories to their five-year plan, they came up empty handed. There was not another man in the country that was willing to go on the record for that. The scientific experts who had devoted their lives to cancer research doubted that a cure could be found anytime soon, even though progress was being made. Certainly, a cure would not be found in ten years. Absolutely not in five! While the majority of cancer experts were happy to endorse an expanded research effort funded by the federal government—free money, yes!—they were guarded about a dash for goal that had been thrown up overnight by a team of three: a zealous philanthropist with shark's teeth and a biting tongue, a pharmacologist with an undistinguished resume and an almost laughable blueprint, and one highly successful oncologist from Harvard University who was still in very early days with his success in treating children with leukemia and had no apparent success in treating adults with solid tumors, if he had any experience there at all.

What Farber and, to a much greater extent, Garb failed to appreciate—and what Lasker had no way of appreciating, given that her degree was in art history—was that cancer is not the moon. It is not made of dust. It does not hold to a fixed and predictable orbit. Cancer presents not one, but a million faces—and all at once. When President Kennedy proposed in 1961 that the United States "put a man on the moon and bring him safely home by the end of the decade," NASA's scientists didn't have to discover the law of gravity. Or figure out the ideal gas law. Or understand that they would need to adjust for the fact that space is curved, or that light is bent by planets and other stars. The laws of physics required for getting to the moon and back were known. The laws of cancer were not. Cancer cells, unlike the moon, are evolving, mutating, moving targets that we still don't fully understand. Cancer experts knew that then. They knew what they didn't know, and they were concerned about a national commitment to cure cancer in five years.

Cure for Cancer came out when the NCI's largest study section was devoted to tumor viruses. Garb, however, was not an enthusiastic supporter of tumor virus research. In chapter 13 of his book, entitled "Search for a Vaccine against Cancer," he concedes to the popularity of tumor virus research, saying, "The search for a cancer vaccine has caught and held the attention of many persons, including some scientists, numerous reporters, and a sizeable segment of the general public. A large and growing portion of our national cancer research program is concerned with the search for viral causes of cancer."[8] Garb indicated that he understood the hope that underpinned tumor virus research: "The prevention of disease by means of a vaccine is about the simplest, least distressing, and most economical prophylactic measure available to the medical profession. If an effective vaccine for cancer could be developed within a reasonable period of time, it would be of the greatest value."[9] He continued, "More than a dozen different viruses have already been found to cause cancer in mice and they show a bewildering variety of behaviors."[10] He allowed, "It might be argued that the fundamental cause of cancer is always a virus, and that other carcinogenic agents merely activate the virus or lower the host's resistance to it."[11] But then without saying why or offering any documentation to support his claim, he said, "The difficulty in making vaccines against all of these different strains would be very great."[12] He did not elaborate or otherwise justify this opinion, leaving the reader to accept it as received wisdom.

But then, Garb went back and erased much of what he had written on the board in support of tumor virus research. He pointed out that scientists had not yet proven that any virus caused cancer in humans. This seemed to suggest that although a great effort had been made to isolate a human tumor virus, nothing had turned up, so there was little more to know. Then, as if to divert the reader's attention elsewhere, he said, "There is convincing positive evidence that many human cancers are caused by agents other than viruses."[13] Garb concluded by saying that even if a tiny fraction of human cancers were one day found to be caused by viruses, the time required to develop effective vaccines would take at least four generations to prepare, thus making them pipe dreams a hundred years away. Garb provided 164 references at the end of his chapter torpedoing tumor virus research, which, as I mentioned previously, was the largest study section at the NCI at the time his book came out. Not one of these references addressed the subject of cancer vaccine development, its difficulties, challenges, or prospects.

It's fair to ask what exactly Garb understood about vaccine development or how he came to his abrupt conclusions annihilating their potential. More importantly, one wonders what Garb understood about tumor virus research, in general. Standing out glaringly in Garb's reference list are twenty-one articles that were published between 1958 and 1966 about what was thought to be a tumor virus responsible for causing lymphoma in children in Africa. Far from being a remote possibility, the authors of these papers were getting very close to nailing it: the *first human tumor virus*.

This is what was actually going on in the references at the back of Garb's chapter about tumor virus research. And obviously, he didn't understand this material. The story goes like this. In 1958, Denis Parsons Burkitt, MD, a surgeon serving in Africa during World War II who would later open a medical practice in Uganda, described a very aggressive, fast-growing malignancy in children. Because Burkitt found that this disease tended to occur primarily in hot, rainy regions of Central Africa, he hypothesized that it might be caused by a virus. Note, this was 1958, about the same time that Ludwik Gross and others began looking for a leukemia virus in children. In 1961, Burkitt made a trip home to England and presented his findings at a scientific meeting. Sitting in the audience was Sir Anthony Epstein, MD, DSc, PhD, an experimental pathologist with an interest in cancer who was hell-bent on being the first man to discover a tumor virus in humans. Epstein also happened to have access to an electron microscope, which proved to be an invaluable asset. Epstein introduced himself to Burkitt, and the two men met for tea after the lecture. They got to work immediately, aided by the help of an assistant, Yvonne Barr. A few years of frustrating work followed, of course, until one day they found what they'd been looking for: cancer cells filled with tumor viruses. They confirmed their discovery in 1965 and called it Epstein-Barr virus (EBV)—three years before Garb's book was published. (Garb includes a reference to this discovery in his book, by the way.)

In 1976—the year Lasker promised that cancer would be cured—the proof was pouring in showing that EBV caused Burkitt's lymphoma—*the most aggressive tumor known to man.* Werner and Gertrude Henle working at the Children's Hospital of Philadelphia began a collaboration with Epstein, Barr, and Burkitt and developed a blood test to screen for evidence of the virus in children with lymphoma. Three years later, in 1979, the proof was in: EBV was found in all patients with Burkitt's lymphoma. EBV was found to transform healthy human blood cells, and the virus was seen to be unique. The paper that planted the flag on the first human tumor virus said, "These findings strongly support a causal relationship between Epstein-Barr virus and Burkitt's lymphoma."[14]

The deadline that Lasker had set for finding a cure for cancer turned out to be the year that Burkitt, Epstein, Barr, Henle, and others began providing irrefutable proof of the existence of the first of many human tumor viruses. It hadn't taken a hundred years; it had taken twenty, start to finish. I have no reason to think that Garb provided references in his book that he did not read. I have to believe that he read the twenty-one scientific papers he referenced about Burkitt's lymphoma and the Epstein-Barr virus in his chapter on tumor viruses and preventive cancer vaccines. But I have every reason to believe that he didn't understand the importance of what he was reading, which is forgivable considering his background. Nor did he comprehend how close Epstein, Barr, Burkitt, and Henle were to proving that tumor viruses cause cancer in humans, not just in mice and monkeys, cats and dogs. By the way, Garb referenced Bittner's work with the mouse mammary tumor virus, but only mentioned it in passing.

To scare off anyone that might still hold out hope that tumor virus research was potentially worthwhile, Garb closed with a note of hysteria: "The vaccine would enhance the cancer rather than protect against it."[15] Bringing this misbegotten diatribe to a close, he said, "Even if such a vaccine were found tomorrow, there is very little chance of its helping anyone living today. It would seem unwise to devote too great a proportion of our resources and energies to this particular approach."

Garb was wrong. Dead wrong. Today scientists believe that 15–20 percent of all human cancers are caused by tumor viruses, an estimate that is far more likely to rise than fall. We can begin with the most common tumor in the world, hepatoma, cancer of the liver. The hepatitis virus causes that, and we have a vaccine to prevent it. The human papilloma virus causes cervical cancer. We have a vaccine to prevent that one too. The human immunodeficiency virus (HIV) causes several cancers, usually in collaboration with other tumor viruses. Today, as a result of the widespread use of antiretroviral therapy in treating patients infected with HIV so that they do not develop full-blown AIDS, HIV-associated cancers are also held in check.

No doubt, Garb had every good intention when he wrote *Cure for Cancer*. His enthusiasm is understandable. His reference to the space program was handy and forgivable. But his ignorant, scornful lashing of tumor virus research was wrong and harmful. As a result of his castigating analysis, the tide began to go out on tumor virus research. Farber, who was pleased and busy with the great work he was doing in treating children with leukemia, apparently didn't appreciate the value of tumor virus research, or he simply chose to look the other way. Lasker, who had no way of assessing the true value of anything related to cancer or tumor virus research, relied on others for her judgment, one who was just plain wrong and one who wasn't paying attention. She, too, can be forgiven; after all, science is not Chagall.

These three musketeers at the center of the Citizens Committee for the Conquest of Cancer thought little of tumor virus research, if they thought of it at all. They had decided that the country needed to go to the moon to find a cure for cancer. Given their ignorance and their arrogance, it was unlikely that tumor virus research would make the list as cargo for the bay.

Chapter 11

The Congress of the United States

Senator Yarborough lost the election to Senator George H. W. Bush, who would later become America's forty-first president. This was a crushing defeat for Yarborough. Lasker described him as "very, very pathetic."[1] But he had come through for her. He was able to arrange for the Senate committee on which he served to establish a panel to evaluate the proposal to create a national effort to cure cancer in five years. Lasker gave him a list of details of what she was looking for, the most important being the creation of a separate cancer agency removed from the direct control of the National Institutes of Health and answerable directly to the president, which would have entirely separate funding by Congress with an appropriation of $400 million, $500 million, and $600 million over the first three years.

On December 4, 1970, the Senate panel issued its report, a blueprint that duplicated Lasker's specifications. The *New York Times* reported the news on page 20 that day, creating the impression in Lasker's mind that its editors weren't taking the proposal seriously enough.

Because Yarborough had lost the election, the conquest of cancer bill that he had introduced in the Senate had to be picked up by someone else. Senator Edward Kennedy obliged and reintroduced the bill as S.1828. Lasker wanted to make sure that the Senate bill passed "overwhelmingly," which meant it would have to be embraced as a bipartisan effort. On February 11, 1971, she hosted a luncheon for thirty-six senators so that she could persuade them of the importance of this gigantic, grid-altering legislation.

Dr. Sidney Farber and others of Lasker's friends addressed the audience with a script she had written that they had rehearsed beforehand. To drive her point home on an emotional level, she had one woman speak "about the human suffering from cancer." Senators Kennedy and Hubert Humphrey came up to her after lunch and told her they were worried that this wildly expensive crash program she was asking for might be a hard sell and telling her, "This is going to take a lot of publicity." Lasker knew that already. She also knew that there would be opposition to it and she "was nervous to know where the opposition would be coming from."[2] For instance, she was a little worried about how serious President Nixon was about getting fully behind the bill. In his inaugural address, January 22, 1971, he had said he wanted to cure cancer in five years, but he only asked for $100 million instead of the $400 million suggested in the panel's report. Lasker took that as a worrisome sign that he might be waffling.

Lasker was right to worry. On February 18, 1971, Nixon announced a Cancer Conquest Program, but he intended to keep it within the National Institutes of Health (NIH) instead of making it a separate agency. He also said that its director would answer to the director of the NIH, which completely undermined Lasker's wishes to have a separate agency that she and other lay activists could control, indirectly, of course. She figured, accurately, that the director of the NIH had gotten to Nixon and convinced him that a moon shot for cancer might fly, but removing the NCI from the NIH would not. Lasker said, "I was very alarmed by this and I realized we were in real trouble."[3]

Lasker went on the attack. She was the honorary chairman of the American Cancer Society, and she was also a member of its board and its executive committee. She appealed to the board for help: "I got the ACS to pass a resolution supporting the Senate bill," which was a duplicate of her original blueprint for a separate cancer agency with independent funding. But even within the ACS she ran into trouble. The ACS's comptroller and some other members of the organization were opposed to a federal crash program for cancer because "people might donate less to the ACS if they knew the government was taking the lead." She pushed harder: "We finally forced the subject on the board of directors and they passed a resolution supporting it and praising President Nixon for his announcement of January 22, 1971."[4] Afterward, she persuaded the ACS to publicize her moon shot for cancer through their chapters, and then she picked up the phone and called Ann Landers.

Ann Landers wrote a weekly column of advice and opinion that was carried in 740 newspapers around the country. She appealed to Landers to help her push her bill through Congress. Landers obliged by writing a "magnificent article" asking her readers to start a "massive letter-writing campaign" to their senators telling them to vote for the Kennedy bill. Nearly a million letters arrived on Capitol Hill, annoying secretaries no end. Two senators received sixty thousand each. Other senators received, on average, twenty thousand letters—in the days when each had to be opened, read, sorted, and answered!

The letter-writing blitz worked beautifully. The Senate, under the guidance of Senator Edward Kennedy, who had replaced Yarborough on the Senate Health Appropriations Committee, passed the bill that Lasker wanted 79–1 and sent it to the House of Representatives for their approval. The one dissenting vote came from Senator Gaylord Nelson. Lasker was particularly peeved about his failure to make her bill unanimous. She had contributed to his election campaign and couldn't understand why he was "hostile" to a moon shot for cancer. "He was really perverse about it. He was led into it by a girl in his office and by a very dumb scientist at the University of Wisconsin who was really offended by the thought that you could ever conquer cancer."[5] In any case, Lasker knew that just because the bill had passed overwhelmingly in the Senate did not mean it would fare as well once it got to the House. First of all, she believed that "in the House they hate the Senators," and secondly, "the people in the NIH were still poking up the scientists to complain and oppose the bill." Lasker realized that she still had a great deal of work to do behind the scenes.

Lasker was adamant that the new cancer agency be established completely apart from the NIH. But the death knell for that came when Representative Paul G. Rogers was appointed as chairman of the House subcommittee that was going to consider the Senate bill for the conquest of cancer. Representative Rogers was not about to be bullied about by Lasker, and he was firmly opposed to changing anything about the relationship of the NCI to the NIH. As Lasker looked back on the events of that year, she regretted not bringing Rogers into the tent earlier. She didn't think he was important or critical enough to her strategy until it was too late to do much about it. Rogers made it clear that he intended to have a long and open discussion about the bill, which meant he would call any witness who would support his point of view, undermining the overwhelming approval that the bill had received in the Senate.

When Lasker got wind that Rogers was against her, she immediately came to the conclusion that "he just didn't realize that the NIH has been philosophically totally opposed to bringing any scientific answers to people. They just want to be a storehouse of information and basic research but not to bring anything to people."[6] Lasker considered the NIH to be a petrified forest of scientific bureaucrats who didn't feel it was "their mission to really go and prove that anything would conquer anything."[7] Recognizing that Rogers might slam the door on creating a separate cancer agency and, thereby, slam the door on her, she met with him in July 1971 to give him her testimony in private. Lasker said that she "gave him an earful," telling him that "the people in the NIH were perverse," that it "never got anything down to the people so that people live longer in better condition," and that the philosophy of its director was "not to bring medical science down to the patients."[8] Rogers listened to her patiently and then sent her on her way.

Opposition to the Senate bill began to mount on many fronts. The NIH and the NCI reached out to everyone they knew to ask them to help destroy the provision that would excise the NCI from the NIH. The response they received was overwhelmingly supportive of keeping the two institutions untouched. The American Medical Association (AMA), dozens of medical societies, and the deans of the medical schools came to the aid of the NIH and the NCI. Lasker's response to this huge surge of resistance to the Senate bill was that "the deans of medical colleges are not in the least anxious to do anything. They don't care whether anything is solved or not as long as they have money to keep the medical schools going." She added that reducing mortality and disability from cancer are "considered ridiculous by deans. They never think of things like that." As for the AMA, she said "they never think about the people who die every year" from cancer. "They don't like it if they have to learn something new."

And then she lashed out at doctors as a group yet again: "Doctors are really quite unimaginative and undynamic on the whole about what to do for people."[9] Lasker's response to this push back was to pay to have the ACS run ads on behalf of the Senate bill in all Congressional districts. It really didn't do much good.

As promised, the Rogers subcommittee hearings turned into an extravaganza in September 1971, not seen since the Circus Maximus in ancient Rome: the only thing missing was Caligula's horse. Senator Gaylord Nelson, the only man who'd voted against S.1828 in the Senate, was invited to appear before Rogers's subcommittee to tell the world why his was the lone opposing voice. Lasker, sitting in the orchestra seats near the front of the room, was not pleased to hear Nelson describe the bill as a "fundamental if not fatal assault on the organizational structure of the NIH." Nelson made it clear that, with the exception of the Laskerites

(Farber, Garb, the American Cancer Society, et al.), no one wanted a separate cancer agency—not the president of the United States, not his science advisor, not the secretary of health, education, and welfare (HEW), and not the entire research community.

Dr. Carl Baker, director of the National Cancer Institute, was invited to testify. Among other things, Baker discussed the importance of research on tumor viruses—a particular interest of his, as he was responsible for creating the Special Virus Cancer Program at the NCI and had overseen work on, among other things, the breast cancer virus.

Dr. James Holland, chief of cancer research at Roswell Park Cancer Institute, also appeared before the committee and spoke about the importance of research on tumor viruses, particularly the breast cancer virus—his specialty.

Secretary of HEW Elliot Richardson appeared before the Rogers subcommittee to explain exactly how a new, independent cancer authority could remain autonomous if it remained within the NIH, a compromise that he hoped would satisfy Lasker. Richardson also mentioned the progress that was being made in the area of tumor virus research.

Henry Kohn, MD, of Harvard Medical School told the committee that any progress made in finding the causes of or cures for cancer would not come as the result of whiteboarding a new organizational chart for the National Cancer Institute but because of the support provided to "the men who carry out the investigations."

Salvador Luria, MD, a Nobel laureate and a professor at Massachusetts Institute of Technology, appeared to say that the world was not yet ready for a moon shot for cancer. It was "self-delusion" to think that a cure for cancer could be found in three to five years, and terribly "misleading of the public" to suggest, let alone promise, something so patently undeliverable.[10]

George Nichols Jr., MD, scientific director of the Cancer Research Institute at the New England Deaconess Hospital, addressed the subcommittee, saying, "We are still today as much in the dark regarding [cancer's] ultimate cause or causes as we were fifty years ago when Peyton Rous first showed that a virus was the cause of a particular malignant tumor in chickens."

After three days of grueling hearings, Congressman Rogers and everyone on his subcommittee left Washington for a day to visit the Roswell Park Cancer Institute in Buffalo, New York. Dr. James Holland, the man who would present the first irrefutable evidence of a human breast cancer virus (in 2006, at the San Antonio Breast Cancer Symposium), was then director of cancer research at Roswell Park. He had already testified before the Rogers subcommittee about the importance of tumor virus research. Roswell Park was a model of how basic scientific cancer research could be integrated across various fields of inquiry and immediately applied to patient care.

It was no easy feat, but the committee produced a cancer bill that was approved by the full House. It was then sent into conference where the Senate and the House could hammer out a final piece of legislation. The result was a bill that kept the NCI intact and within the NIH. A cancer advisory board was created that would oversee the national effort to cure cancer in five years, but, importantly, the board would not have veto power over specific research projects set forth in the program. The board would have eighteen members, with six lay citizens, nine doctors, the director of the NIH, the secretary of HEW, and a chairman. The appropriation was set at $1.6 billion for the first three years. In the end, all that Lasker really got of what she wanted was a huge increase in the funding for cancer research, primarily directed toward diagnosis and treatment rather than what she also referred to as "so-called basic research."[11] She was not pleased and felt somewhat resentful about the outcome. "If it hadn't been for me there wouldn't have been any NIH as it now exists, nor any money available," she said.[12]

On December 23, 1971, President Nixon invited 250 people to the White House to watch him sign the National Cancer Act into law. The new president of the American Cancer Society was there to take the pen from Nixon and later appeared on television to support the bill: if you can't fight it, feature it. Lasker didn't think there was any need to discuss the future of cancer research beyond 1980, for she was convinced a cure would be found within the decade. She sat in the Senate and then in the House during the committee hearings in the run up to the National Cancer Act, but as always, she refused to testify in public "because the trouble with that is that you're taking the clothes off the boys."[13]

Benno Schmidt, an attorney and political operative who was also chairman of the board of Memorial Hospital in New York, became the head of the President's Cancer Panel, a three-man group at the NCI whose job it was to keep the president informed about the progress that was being made in finding a cure for cancer in time for the nation's bicentennial. He was a man who liked to see results. In fact, he demanded them.

Dr. Baker, director of the National Cancer Institute, and the man who had helped to build and lead the Virus Cancer Program at the NIH, was fired.

Dr. Robert Marston, director of the National Institutes of Health, was fired. His replacement was Robert Stone, MD, a man without research expertise from the University of New Mexico, who was said to be a skilled manager. Apparently, he wasn't skilled enough because he was also fired three years later.

Other important leaders at the National Cancer Institute and the NIH stepped down or stepped away. Robert Berliner, MD, deputy director for science, left to become dean of the Yale Medical School. John Sherman, MD, deputy director of the NIH, left discouraged and disillusioned.

Frank Rauscher Jr., MD, a pioneer in tumor virus research at the NCI for seventeen years, and the man in charge of the program aimed at understanding the causes of cancer, resigned when it became clear that Benno Schmidt was running the show.

Umberto Saffiotti, MD, head of the NCI's carcinogenesis investigation unit, resigned in a loud flurry of frustration when his funding and staff were cut to shreds.

***

The preceding summary of the enactment of the National Cancer Act of 1971 is but a brief account of one of the most intriguing episodes in American history. For a more thorough analysis of the events that led up to this enormous piece of legislation, I highly recommend R. A. Rettig's splendid book, *Cancer Crusade* (see Bibliography). It would make excellent reading on a dark and stormy night. On the other hand, if you're looking to be amused, then search the secondhand book market for Solomon Garb's book, *Cure for Cancer: A National Goal* (see Bibliography).

Chapter 12

Zinder and zur Hausen

Dr. Carl Baker was a physician who developed an interest in research. He began his career at the National Cancer Institute in the mid-1950s, fully expecting to remain in the laboratory for the rest of his career. But then he developed an allergy to animals that he could not overcome. This drove him from the laboratory, which he loved, and into administration, where he excelled.

Baker was highly organized and keen for detail. He was a natural leader and a decent man who could get things done smoothly and efficiently. As a result of these attributes and skills, Baker rose steadily through the ranks of the Public Health Service, at the National Cancer Institute, and in 1970, he became its director.

Baker was instrumental in coordinating tumor virus research at the NCI, which began in 1964 with the creation of the Leukemia Virus Study Section. Shortly thereafter, scientists discovered that viruses were capable of producing solid tumors as well as leukemia and lymphoma. Under Baker's leadership, the NCI committed increasingly more money and manpower to tumor virus research until the Special Virus Cancer Program (SVCP), the study section's last iteration, became the largest study section at the NCI. There was no doubt in Baker's mind, or in the minds of the hundreds of scientists and staff who worked in the Special Virus Cancer Program, that this was smart money well spent. By 1970, the NCI was spending $10 million a year on tumor virus research when the average cost of a home was $24,000.

The National Cancer Act of 1971 provided hundreds of millions of dollars a year to the NCI to find a cure for cancer. The three-man President's Cancer Panel at the NCI reported on the progress that was being made in reaching this goal. The act also provided for a supervisory board (half medical professionals, half lay citizens) whose job was to review and comment upon the NCI's conquest of cancer plan. The National Cancer Advisory Board consisted of twenty-three members. Mary Lasker was one of these. As specified in the National Cancer Act, the director of the NCI was required to formulate and present to the National Cancer Advisory Board its National Cancer Plan, a blueprint for ending cancer within five years. To prepare for his first presentation of the National Cancer Plan to the National Cancer Advisory Board, Dr. Baker studied NASA's strategic plan for the space program. He adopted what he thought might be advantageous for the moon shot for cancer and then recruited 250 scientists from across the country to assist him in drafting the National Cancer Promise, the official contract to fulfill the mission set forth in the National Cancer Act.

The scientists were assigned to one of forty panels, which were organized by topic and objective, and then they met for a three-day conference that took place in the quiet Virginia countryside (close enough to Washington to be convenient for travel, but far enough away that they were not distracted by the din of beltway buzz). To meet the goal of finding a cure in five years, the National Cancer Plan was forced to focus on diagnosis and treatment rather than on basic research or prevention of disease. A nod was made in the direction of smoking cessation. Words were spoken about the need to address environmental carcinogens. But the tie and rail to cure were laid down on the field of categorical research, not because the scientists thought it was the best course, but because, under pressure as they were, they had little choice.

Where did tumor virus research, now in full bloom at the NCI, stand in the midst of the National Cancer Plan? It stood at the gallows. There was a small but growing opposition to tumor virus research arising on the periphery of the NCI. First, there were those outside of the NCI who thought it worthwhile and wanted a larger piece of the funding pie to support their laboratories. Second, there were those who wanted the whole thing shut down as a waste of time. So on the one hand, there were disgruntled scientists accusing the Special Virus Cancer Program of hogging all the money for its own research, charging the NCI leadership of running an "old boy network," keeping a "closed shop," and hoarding the millions for a chosen few. On the other hand, there were the men who felt that categorical research had to take precedent if a cure was to be found in five years, making tumor virus research and vaccines as far as way as Pluto—too far a reach. They all came complaining, vehemently, to Benno Schmidt and the other members of the National Cancer Advisory Board, demanding that something be done. Should he spread support around or end it completely? Schmidt considered the matter carefully.

On paper, it certainly appeared that the lion's share of Special Virus Cancer Program funding orbited around a tiny constellation of investigators employed by the National Cancer Institute. However, this was less incestuous than it looked. The pattern of disbursement was less a scheme than a consequence of two separate factors that converged over time to shorten the radius around which the money circulated. In the beginning, few scientists were willing to put their careers on the block of tumor virus research, for as we've seen, it was perilous to do so. Even when Peyton Rous demonstrated, time and again, that the virus he had isolated in chickens (1911) was an infectious agent capable of producing malignant tumors in healthy chickens, doctors looked him straight in the eye and told him that his tumors were *not cancer*. The scathing humiliations inflicted upon Peyton Rous were sufficient to scare off all but the most intrepid scientists. Only those who found a safe harbor and independent funding could afford the luxury of chasing tumor viruses. It wasn't until scientists at the National Cancer Institute found evidence of a monkey leukemia virus in children with leukemia that the ice began to thaw. But still, it was hard for these men to be taken seriously. It was only when the polio vaccine was finally perfected that scientists working on the polio virus at the National Institutes of Health began to look for something else to do, some other virus to work on. Word got around the campus at the NCI and the polio experts were invited to help with the Leukemia Virus Study Section. Having more hands to the pump really increased the momentum for tumor virus research, and that's when progress really started to be made, but only in and around a vanguard few.

When it became clear that tumor viruses caused solid tumors (like breast and lung tumors) as well as leukemia and lymphoma, research elsewhere began to take root. Time and discoveries did little to change the majority view outside the walls of the NCI, however; and so, the advances that were made in this field were circumscribed around a few intrepid investigators who were, for the most part, holding cover within the National Cancer Institute. A handful of scientists elsewhere began breaking new ground too, but by the time the new converts came on the scene, the small band of investigators who had pioneered tumor virus research at the NCI had moved far enough in their own investigations that they felt the rookies lagged too far behind to warrant a big piece of the federal funding pie. In short, the pioneers were no longer fishing around: they were hot on the trail of high-value targets. Funding for this kind of targeted research was provided for by contracts. Naturally, the rookies were peeved that the NCI wasn't giving them the money they requested to begin their own investigations of tumor viruses. Funding for this type of startup research was provided for by grants. The newly converted wanted more Special Virus Cancer Program money to be given out as grants, to them, and less spent funding projects for the closed shop gang. That made sense to Schmidt: spread it around.

Meanwhile, on the sidelines of this debate were reporters behaving like armchair referees. In 1971, as the National Cancer Act was becoming an avalanche of legislation, Nicholas Wade, a staff writer for *Science*, wrote a scathing editorial criticizing the Special Virus Cancer Program for its lack of progress in proving that viruses cause cancer in humans. Wade, who was a journalist and not a scientist, claimed that the NCI had failed in its efforts to prove the worth of tumor virus research because its investigators were incompetent! He quoted a scientist in academia (who refused to be identified) who claimed that the main problem with the Special Virus Cancer Program could be traced to "second class research" that lacked "enough of an intellectual base." John B. Moloney, PhD, who was then scientific director for viral oncology at the National Cancer Institute, was livid. Moloney said, "Wade presented a biased and unbalanced account of what the Special Virus Cancer Program was trying to do, a feeling shared by many other participants in the Program." The War on Cancer had become a war, indeed.

The second factor that contributed to the appearance of a "closed shop" funding operation among a small group of investigators in the Special Virus Cancer Program resulted from the National Cancer Institute's desire to fund tumor virus research in a cost-effective manner. Let me explain. Scientists are always in need of material and supplies to conduct their experiments: animals of every type and size, chemicals, supplies for growing tumor viruses in culture, and so on. Scientists need a lot of stuff, and they need a steady supply of it. And importantly, it has to be *standardized stuff*. These materials and supplies have to be standardized so that the results of the experiments are reproducible. Standardization of resources was critically important for scientists in the Special Virus Cancer Program because tumor viruses are tricky enough as it is. The bottom line was to standardize as much "stuff" as possible to make the experiments as reproducible as possible with the hope of gaining as much credibility as possible in order to make progress and grow the study section.

Faced with the daunting task of obtaining a vast, ready, and reliable supply of standardized material to meet the needs of tumor virus research, administrators of the Special Virus Cancer Program had a choice to make: either manufacture all this stuff themselves, on the campus of the National Cancer Institute, or contract the work out to other institutions in much the same way that the Department of Defense contracts out the building of fighter jets to Boeing. Obviously, it made more sense to enter into contracts with companies like Pfizer to supply the resources needed for tumor virus research.

As a consequence of a decision based solely on considerations of cost and efficiency, it looked like the money that was being allocated to contracts was boomeranging straight back to the men who ran the laboratories at the NCI where most of the tumor virus research was being done. To anyone trying to follow the money and connect the dots at the Special Virus Cancer Program, it looked like a tightly orchestrated, restricted network—very "old boy." That's what it looked like to Schmidt, too.

While there was no doubt that Schmidt was a capable executive, he was totally incapable of rendering an opinion as to the scientific value of tumor virus research. For this, he had to rely on the advice of others.

With the appearance, at least on paper, of "insider trading" within Special Virus Cancer Program, and the strident accusations of investigators who felt they'd been excluded from the NCI's money train, Schmidt and the members of the National Cancer Advisory Board punted the ball as far down field as possible. In March 1973, the board created a committee to investigate the Special Virus Cancer Program. Norton Zinder, PhD, of Rockefeller University, where Peyton Rous had discovered the first tumor virus in 1911, was asked to chair the committee. Oddly enough, the members of the Zinder Committee, as it came to be known, were selected specifically because *they lacked expertise in tumor virus research*. It was hoped that having nonexperts review the work of the Special Virus Cancer Program would serve to minimize bias in rendering an opinion about the value and integrity of the work that was being done. But this decision, well intentioned though it may have been, came at the cost of being able to appreciate what progress had been made and its potential for the future control of cancer.

While the Zinder Committee was mulling over the Special Virus Cancer Program, Mary Lasker, the only woman on the National Cancer Advisory Board, was still fighting a battle that she had lost but had not conceded. On March 20, 1972, two hundred people squeezed into a conference room in a hotel in rural Virginia for Baker's formal presentation of the NCI's National Cancer Plan. The thirty-three members of the National Cancer Advisory Board were hunched next to Baker and Robert Marston, MD, director of the National Institutes of Health, on a raised platform for better viewing by the assembled multitude. Everyone got a chance to introduce themselves and say a thing or two; opening remarks and introductions took two and a half hours.

After a break for coffee, Baker stepped forward to unveil the plan. Recall that, originally, when Mary Lasker submitted her plan for the National Cancer Act to Senator Yarborough, she had insisted that the National Cancer Advisory Board be given the power to approve every research project of the National Cancer Institute. This was one of the first things erased as the bill moved through Congress: the scientists would have none of it! This provision was dropped early on, well before the final bill for the National Cancer Act came forward for a vote. But Mary Lasker was not one to take no for an answer. According to "An Administrative History of the National Cancer Institute's Viruses and Cancer Programs, 1950–1972," written by Carl G. Baker, MD, and published by the National Institutes of Health, which is retained in the archives of the National Library of Medicine, Carl Baker began presenting the first item in the National Cancer Plan. He had no chance to move on to explaining the second item when the following scene took place. (Benno Schmidt, chairman of the National Cancer Advisory Board, was also on stage officiating for the day.)

> During presentation of the contract-supported activities, some new members raised questions about the use of contracts [funding of specific items such as laboratory animals] and wondered if the money going for funding of contracts might be better spent on grants [funding individual investigators to support their research]. These comments were not unusual from those without experience with contracts. Dr. Baker again pointed out the need for funding both basic exploratory research (grants) and multidisciplinary targeted research and developmental efforts, as well as development and distribution of defined resources (contracts). When a particular contract proposal was being discussed, Mary Lasker made a motion to approve the individual contract proposal, thus again attempting to establish that approval of individual contracts would require Board approval. Attempting to avoid the setting of precedent, Dr. Baker stated that he did not need this recommendation from the Board to fund the specific contract. Mr. Schmidt tactfully reminded the group that the new Act did not allow this action to be one the Board could take.[1]

Checkmate, Mary Lasker.

In April 1974, the Zinder Committee issued its report to the National Cancer Advisory Board. While the report fully endorsed tumor virus research, per se, it was strenuously opposed to any effort that might direct this research toward the development of preventive cancer vaccines. The Zinder Committee had essentially come to the same conclusion as Solomon Garb: because after decades of research scientists had not yet made a breakthrough linking a virus to human cancer, the funding of cancer vaccines was deemed to be a waste of time and money. If the goal of tumor virus research was to eventually develop a preventive cancer vaccine, and development of cancer vaccines was off the table, then what was the point of funding the research in the first place? Without intending to do so, the Zinder Committee had torpedoed tumor virus research by making it relatively pointless, at least as far as the race for a cure was concerned.

The Zinder Committee's verdict against tumor virus research as prelude to the development of preventive cancer vaccines came at exactly the wrong moment. Not only were Burkitt, Epstein, Barr, Henle, and others drilling down on proof of the existence of the first demonstrable human tumor virus (the one that causes lymphoma in children), but Harald zur Hausen was on the threshold of finding the second one: the human papillomavirus (HPV). The year after the Zinder Committee essentially took the ax to tumor virus research, zur Hausen discovered that the human papillomavirus was the cause of cervical cancer in women. Unfortunately, by then it was too late. It didn't matter how strong zur Hausen's data were, how irrefutable his experimental results, how compelling his epidemiological findings: no one was interested. The case had been settled and was closed.

After putting the kibosh on cancer vaccines, the Zinder Committee took the opportunity to elaborate on what it felt were the most significant failings of the Special Virus Cancer Program. It agreed with the popular myth among competing academics that the program investigators were running a closed shop designed to fund a small, exclusive group of scientists at the National Cancer Institute. Then in a recapitulation of Nicholas Wade's diatribe that was published in *Science*, the Zinder Committee called the scientists running the Special Virus Cancer Program inept: scientists at other institutions could have done a much better job of finding the link between viruses and human cancers.

The report continued: "We are quite sure that much more would have been accomplished if equal support had been provided on a competitive basis to many more laboratories with greater capability and experience in particular areas."[1]

Because it was felt that the Special Virus Cancer Program had failed to distribute funds for tumor virus research more equitably to other institutions, and because its scientists weren't up to the task of making clinically useful discoveries, the Zinder Committee recommended that another committee be created to scrutinize, line by line, every grant and project funded by the Special Virus Cancer Program. This was especially insulting, particularly to Baker, who had grown it and run it (beautifully) for years. You see, the National Cancer Institute maintained a strict policy of peer review of every research grant and project. The leadership at the NCI and the SVCP had already evaluated each project, line by line with a fine-toothed comb, before settling on the most promising. The Zinder Committee did not feel that was enough scrutiny. Now, a new panel of "experts" was to be assembled to render its judgment of what was worth funding and what was not. The need for a guillotine, though not specifically stated, was implied.

Harold Amos, PhD, of Harvard University was asked to chair the new committee. The Amos Committee, as it came to be called, reviewed every grant and project of the SVCP and found only half worthy of continued funding, but they didn't hold that worth in very high esteem. As a result, the Amos Committee thought it best to let every research project expire when its funding ran out and then revisit each one individually to determine which should be renewed. The hope was to expand the SVCP beyond the walls of the NCI, steering it into more competent hands, and distributing the money more equitably in the process. And while the Amos Committee did not intend to annihilate tumor virus research altogether, that's exactly what happened. As contracts and grants in the SVCP expired, the money set aside for tumor virus research was not redistributed to outside investigators at all; it was gobbled up and used to fund other things like the newly discovered oncogenes.

In discussing the history of tumor virus research at the NCI, particularly the untoward consequences of the Zinder and Amos Committees,

Robert Gallo, MD, a virologist who codiscovered HIV, recalled in his oral history: "There was a big push to get rid of the Virus Cancer Program. There was a big push to . . . just forget all the virus work."[2] The big push worked. The Zinder Committee's push ended tumor virus research at the NCI and just about everywhere else for a very long time. Once the NCI had to abandon tumor virus research when its funding was no longer renewed, other institutions followed suit. Samuel Herman, DDS, PhD, a scientist who was working at the NCI during the demise of the SVCP, grimly recalled, "You begin to see what are the implications of lay people making political decisions in science."[3]

The mortal wound for tumor virus research came at the hand of National Academy of Sciences when, in 1974, it issued a dire warning about the potential risk tumor viruses posed to the public. It called for a moratorium on all tumor virus research until scientists could guarantee that they would not accidentally spread cancer contagion around the world. And so, with the rug pulled from under the Special Virus Cancer Program by the Amos and Zinder Committees and the red light issued by the National Academy of Sciences, the roof caved in completely on tumor virus research. Scientists who'd devoted their entire lives to tumor viruses scrambled for safer shores. Harold Varmus, MD, who was working at the NCI when the National Cancer Act got under way—and was its director until March 2015—discovered the first oncogene while he was working with John Bittner's breast cancer virus. Because tumor virus research had paved the way for the discovery of oncogenes, and because it was hoped that being able to switch oncogenes "off" would be the key to curing cancer, many tumor virus experts left their empty laboratories and moved to where the oncogenes were thriving.

Dr. Carl Baker, who was appointed the director of the National Cancer Institute in 1970, lost his job in 1972, a year after laying out a strategic plan for the War on Cancer that included funding tumor virus research. Two years later, the National Cancer Advisory Board was having problems of its own. Its members couldn't agree about the best approach. The *Wall Street Journal* reported that a sizable portion of its members were ready to resign in frustration. Lasker didn't think it was a very good board "because we have James Watson on it, which is the famous scientist—the double helix—but he talks endlessly and not to any point in the meetings."[4]

The year that the Zinder Committee issued its report about the Special Virus Cancer Program (1974), Dr. Harald zur Hausen, a German doctor with an interest in tumor viruses, presented his research on the HPV at a conference in Florida. He was sure that HPV was the cause of human cervical cancer. He had made the "giant leap for mankind" that all tumor virus researchers had hoped to achieve, that of finding the first human cancer virus. He presented his evidence. He showed his slides. He discussed his data and used it to justify his assertion that HPV caused cervical cancer. The audience dismissed every bit of it. They said his data were erroneous, his investigations were irrelevant, and his conclusions were unsupportable; he hadn't found a thing worthwhile. Undaunted, zur Hausen continued his research into HPV at the German Cancer Research Center at the University of Heidelberg. In the course of his investigations, zur Hausen discovered that two strains of HPV (16 and 18) caused 90 percent of human cervical cancer. In 1984, he approached drug companies with his discovery and encouraged them to develop a vaccine against HPV. "But the companies I approached did not believe that this would be profitable and said there were more urgent problems to be solved."[5] By 1991, additional studies confirmed that HPV was the primary cause of human cervical cancer. Fifteen years later (2006), the world's first preventive cancer vaccine was released: Gardasil. It targets HPV so that women do not become infected with the virus and, therefore, do not get cervical cancer. It is 100 percent effective and will be the means of eradicating this disease worldwide.

In 2008, zur Hausen was awarded the Nobel Prize in Medicine. In his Nobel Lecture, zur Hausen said, "Slightly more than 20 percent of the global cancer burden can currently be linked to infectious agents, including viruses, bacteria and parasites."[6]

Zur Hausen launched into a discussion of other possible human tumor viruses, starting with the mammary tumor virus. "Is it possible that a similar mechanism contributes to human mammary cancer? A few data seem to support this notion," he said. He concluded, "Human breast cancer remains an interesting candidate for a viral etiology."[7]

Chapter 13

Moore and Ogawa

The story of the breast cancer virus might have ended with the Zinder and Amos Committees having the final word. It might have ended when the last remaining project in the Special Virus Cancer Program eventually expired. But mercifully, it did not. The science was too compelling. The data were too alluring. Curiosity would not let go of the dangling, unanswered question, *Does a virus cause breast cancer in women?* Without knowing it, zur Hausen helped to keep it all alive. By hammering away at the human papillomavirus, and then proving that it caused cervical cancer in women, he rescued tumor virus research from oblivion—even if the majority of the medical profession had left it to rot and die and decompose as fertilizer for oncogenes and other things. A few scientists, working like hermits in the desert and living on the equivalent of locusts and what honey they could find, were equally determined to pursue the prey they thought was pursuing them: tumor viruses such as Epstein-Barr, hepatitis, and the breast cancer virus.

In the 1950s, scientists working with the mouse mammary tumor virus (MMTV) developed a way to grow mouse kidney cells in the laboratory and then infect them with the virus. They were interested to see what effect, if any, the introduction of MMTV would have on the growth of these living kidney cells. Three Russian scientists discovered that when mouse kidney cells were infected with MMTV, they grew more rapidly than normal cells. It was an interesting observation, because one of the most important characteristics of cancer cells is that they grow rapidly. Then a group of American scientists repeated the experiment. And they got the same results: MMTV increased the growth rate of mouse kidney cells. But they saw something else, something that the Russians had not reported. In addition to growing faster, the mouse kidney cells were growing abnormally. Abnormal cells that are growing rapidly are known to be precursors to malignancy. Right before their eyes, the American scientists were seeing normal mouse kidney cells infected with MMTV grow rapidly and abnormally. It was scary to behold, but informative.

Other scientists, following in the footsteps of the Russians and the Americans, repeated the experiments. As each scientist tends to stand upon the shoulders—sometimes the heads—of others, other scientists began to learn new things about MMTV-infected mouse kidney cells. What they learned was even more alarming and surprising: not only did MMTV increase the growth of kidney cells, it accelerated their growth—which is to say, their growth rate continued to increase day by day. It was as if MMTV had its foot on the accelerator and was pressing down relentlessly. But there was more: the cells *lived longer* when they were infected with MMTV. They lived longer and they grew faster over time. It was a frightening observation.

In August 1970, while Mary Lasker's Citizens Committee for the Conquest of Cancer was drawing up its blueprint for the War on Cancer, researchers at Pfizer, a drug company located in New Jersey, found viruses in the breast tumors of rhesus monkeys that were similar to MMTV. This momentous discovery introduced the distinct possibility that scientists would eventually find a similar breast cancer virus in humans. The Pfizer researchers published their findings in a peer-reviewed journal, saying that it was "of great significance and warrants further investigation to determine its role in (human) breast cancer."[1]

The following year, as the details of the National Cancer Act were being hammered out on Capitol Hill, another investigator in New Jersey discovered evidence that more directly linked MMTV to human breast cancer. In the paper he published in 1971, "Search for a Human Breast Cancer Virus," Dan Moore, PhD, of the Institute for Medical Research in Camden, New Jersey, reported that "some human milk contains particles physically identical to the mouse mammary tumor virus. Tumorigenesis in the mouse could serve as a model in the study of human breast cancer."[2] Moore discussed the various factors that contributed to breast cancer in mice: a genetic predisposition, the presence of the mouse mammary tumor virus, and the presence of female hormones required to stimulate growth of the tumors. In finding MMTV-like viruses in human breast milk, Moore felt justified in making the bold statement (with which the editors of *Nature,* the journal that published his paper, must have agreed): "It seems that, as in the mouse, a viral factor is involved in the etiology of human breast cancer. Results of recent studies on human milk [in which viral particles similar to MMTV had been found] strongly support this idea."[3] Moore concluded, "Similarities between cancer of the breast in mice and women are too extensive to be coincidental," and "human breast cancer may also be a viral disease. Recent studies suggest that the human virus may be similar to if not identical to the mouse virus."[4] To emphasize, this discovery was made forty-six years ago, and it was published in *Nature*, one of the most prestigious scientific journals in the world.

The following year, 1972, R. Axel, MD, of the Institute of Cancer Research at Columbia University published a paper, "Presence in Human Breast Cancer of RNA Homologous to Mouse Mammary Tumor Virus RNA." (*Homologous* means "similar." Unlike DNA, RNA can move in and out of the nucleus of the cell to carry genetic messages.) Dr. Axel reported that he had found genetic evidence of MMTV (in other words, RNA) in the tissue taken from women with breast cancer. Moore had found evidence of MMTV in human breast milk in 1971, and one year later Axel had found evidence of MMTV in human breast cancer specimens. Axel's paper was published in *Nature*, too. He came to the same conclusion as Moore the previous year, saying that this discovery provided "the most compelling evidence available for the involvement of virus-related information in human breast cancer."[5]

While the Zinder and Amos Committees were busy directly and indirectly shutting down tumor virus research across the country, the evidence in support of a human breast cancer virus continued to pile up in crevices they could not reach. In 1978, Dr. Stefan Zotter, MD, from the Institute of Pathology in Dresden, Germany, found viruses similar to MMTV in human breast milk, just as Dan Moore had, concluding that this was evidence "of the existence of a MMTV-related human" breast cancer virus.[6]

Again, in 1978, other corroborating evidence of a human breast cancer virus emerged, this time from Japan. Jun Ogawa, PhD, from Kyoto University, reported finding evidence of MMTV in the blood of Japanese women with breast cancer. He discovered that 60 percent of breast cancer patients had antibodies to MMTV circulating in their bloodstream, prima facie evidence that they had been infected with the mouse breast cancer virus. This is extremely important. There is always the question about which species a particular virus is capable of infecting. By 1978, there was no doubt that MMTV infected mice. It also infected cats and monkeys. But when Dr. Ogawa reported finding antibodies to MMTV in the blood of women with breast cancer, it became clear that MMTV could infect humans too. Not only did Ogawa find evidence of antibodies to MMTV in the bloodstream of 60 percent of women with breast cancer, he found that 25 percent of women with a benign tumor called a fibroadenoma were also infected with MMTV. (In some studies, the presence of a fibroadenoma increases the risk for breast cancer.) Normal healthy women, on the other hand, had a much lower rate of infection with MMTV: only 11 percent. The relationship between the infection rates (in other words, 11 percent of healthy women, 25 percent of women with fibroadenoma, and 60 percent of women with breast cancer) is also profoundly important. It implies that as breast tissue becomes increasingly abnormal (from completely benign, to fibroadenoma, to cancer) the incidence of infection with MMTV rises with it.

Ogawa also found something else that was surprising and even more interesting: "In a considerable number of positive cases, the antibody tended to disappear within various lengths of time after surgical operation of the breast cancer."[7] Dr. Ogawa continued, "an increasing number of reports are being published suggesting that MMTV or its related virus is somehow involved in human mammary carcinogenesis."[8]

In 1978, Ricardo Mesa-Tejada, MD, from the Institute of Cancer Research at Columbia University in New York City, found evidence of MMTV in human breast cancers. However, the distribution of MMTV within the specimens was not uniform. In both mice and women, the tumors showed "patchy" areas where MMTV predominated. This suggested that MMTV does not manifest in all cells of the tumor equally—extremely important! This finding is consistent with earlier papers published by scientists at the National Cancer Institute and elsewhere in which it was found that tumor viruses could often be the cause of cancer but remain elusive within the cells. The results of Dr. Mesa-Tejada's work revealed that some breast cancers that resulted from MMTV infection had evidence of the virus, while other cells either hid the evidence or, once infected, began to grow independent of further MMTV stimulation. I realize that in discussing this point, I am adding more details to an increasingly complex picture of how MMTV works to cause breast cancer in animals and, possibly, women; but the fact that MMTV can infect a cell, cause cancer, and then disappear from the scene of the crime is an important discovery to keep in mind as we move forward in this story.

Despite the progress many scientists were making in the 1970s tracking down evidence of a human breast cancer virus similar to MMTV, one soon-to-be famous scientist failed to find anything worth getting excited about. He was important for three reasons: (1) because of where he was working when he failed to find a link between MMTV and human breast cancer, (2) what he found instead, and (3) what happened to the hunt for a human breast cancer virus as a consequence of his new discovery. That man was Harold Varmus. I introduced you to him in the previous chapter. He was the man who tried but couldn't find MMTV in human breast cancer, but found the first human oncogene instead.

Dr. Harold Varmus was raised in a family where it was expected that he would become a doctor, like his father. But Harold wasn't as interested in medicine as literature. He took his undergraduate degree at Amherst and then obtained a master's degree in English literature at Harvard University. Finding that there was little future in literature, he caved in to the family tradition and entered medicine because it "opens all doors." Varmus graduated Columbia University medical school at the height of the war in Vietnam. Like most newly minted physicians at the time, his first priority was to avoid the draft. In 1968, Varmus received a commission in the Public Health Service, which served as a popular substitute for doctors wanting to skirt the draft, and took a job at the National Cancer Institute "seeking two things: the credentials to become a medical school professor and an alternative to service in Viet Nam."[9]

Varmus began working in a laboratory that was investigating tumor viruses and was quickly seduced by his own curiosity. As he described in his commencement address to Harvard University on June 6, 1996, he fell in love with "the sweet anticipation of my own results."[10] (Varmus's mother had died of breast cancer, as had her mother. Quite naturally, he was keen to discover a role for MMTV in human breast cancer. But his research "yielded no insights into human breast cancer."[11] It did, however, pave the way for a breakthrough: the discovery of the first oncogene (a gene whose activation leads to cancer). Varmus abandoned MMTV and headed out to what looked like a far more promising field.

In 1989, Varmus and J. Michael Bishop, MD, were awarded the Nobel Prize for their discovery of the first oncogene. In 2010, Varmus was appointed director of the National Cancer Institute and remained in that position until March 31, 2015. He is no more enthusiastic today about the existence of a human breast cancer virus than he was in 1968. There is a "glass ceiling" for any scientist that takes human mammary tumor virus seriously and wants to pursue its role as a cause of human breast cancer. It is exceedingly difficult for any investigator to promote a hypothesis that the director of the National Cancer Institute just doesn't buy. This is something that's true in every discipline, not just medicine. Over a hundred years ago, it was also very difficult for scientists like Niels Bohr and Erwin Schrödinger to sell the world on quantum mechanics (a tenet of physics predicated on probabilities) when Albert Einstein refused to believe it. Einstein famously said, "God doesn't play dice" with the universe. Einstein was convinced that reality existed apart from observation and probabilities, and the quantum physicists insisted that reality was disturbed by observation and, at the subatomic level, probabilities were about as accurate as one could hope to get. The quantum physicists were right. Einstein was wrong.

In Manjit Kumar's wonderful book, *Quantum* (see Bibliography), the author described Bohr's explanation for Einstein's recalcitrant opinion: it was his "very success, implied Bohr, that kept him anchored to the past."[12]

It can be argued that you can't get much smarter than Einstein. Like Einstein, Varmus won the Nobel Prize. So, yes, winning the Nobel Prize means you're smart, but it doesn't always mean you're right.

Ignorance, politics, vanity, power: they all get in the way.

Chapter 14

Baltimore, Temin, Nusse, Callahan, and Franklin

The electron microscope was developed in the 1940s. It allowed scientists to visualize images much smaller than could be seen with the light microscope. For the first time, structures that were measured in nanometers, like viruses, came into view. One nanometer is one billionth of a meter. There were approximately six thousand different viruses classified in 1950, but in general, they fall into two broad categories: those with genes that are made of DNA, and those with genes that are made of RNA. Scientists discovered that some RNA viruses, such as the mouse mammary tumor virus, are able to have an effect on the host genome, which is made up entirely of DNA. They were puzzled by this finding and couldn't understand how RNA viruses exerted their cancer-causing effect on the DNA genes that resided within the genome.

Then in 1970, David Baltimore, PhD, and Howard Temin, MD, solved the mystery. Dr. Baltimore had fallen in love with biology during a high school summer internship that he spent at Jackson Memorial Laboratory in Bar Harbor, Maine. He settled on science as a career and graduated Swarthmore College in 1960 with a degree in biology. Baltimore earned his doctorate at Rockefeller University. It's easy to see how he might have become interested in tumor viruses considering that he'd spent time in the laboratories where the first two tumor viruses were discovered (the Rous sarcoma virus at Rockefeller University and the mouse mammary tumor virus [MMTV] at JAX Lab). Baltimore had taken a position at MIT and was working there in 1970 when he discovered just how RNA tumor viruses manage to park their genes inside the DNA of the host genome. Dr. Howard Temin, who was working separately on the same problem at the University of Wisconsin, solved the mystery at about the same time, too.

What Baltimore and Temin discovered was that RNA tumor viruses carry a "secret code"—*reverse transcriptase*—a unique enzyme that makes a DNA copy of the RNA tumor virus genes. With the help of this clever trick, reverse transcriptase transforms an RNA virus into a DNA duplicate. Because the RNA virus now masquerades as DNA, it is able to enter the nucleus of the cell where the DNA of the host genome resides and insert its viral genes without causing so much as a whisper of a stir. Baltimore and Temin's discovery of reverse transcriptase was a gigantic breakthrough in helping scientists understand how RNA tumor viruses could transform normal cells. This discovery made them rock stars overnight. Eventually, it earned them the Nobel Prize. The next step, of course, was for scientists to figure out just what those viral genes did after they got into the host genome. How, exactly, did they cause cancer?

Every investigator working on RNA tumor viruses began looking for, and finding, reverse transcriptase in cells infected by these viruses, and in the cancer cells that these viruses produced. Reverse transcriptase was found in breast cancers associated with MMTV, and it was found in dozens of other tumors linked to tumor viruses. Interestingly, it was the discovery of reverse transcriptase that led directly to the discovery of HIV as the cause of AIDS ten years later. HIV is an RNA virus, and it's a tumor virus, too. As it turns out, HIV and MMTV belong to the same category of viruses.

In addition to the two broad categories of DNA and RNA viruses, it was decided that RNA viruses that possessed reverse transcriptase become part of a new class of viruses called *retroviruses*. Once Baltimore and Temin discovered reverse transcriptase, its mere presence in a cell was considered to be prima facie evidence of infection with a retrovirus. The presence of reverse transcriptase in a tumor cell was considered to be evidence of infection with a tumor virus. The next step was always to discover which tumor virus that was. In the case of breast cancer in mice, the retrovirus was always found to be MMTV.

Once scientists discovered that RNA tumor viruses use reverse transcriptase to make DNA copies of their genes, they were curious to discover the exact mechanism by which these copies—called *proviruses*—were inserted into the host genome. Scientists knew that because the copy was made of DNA, it could get into the nucleus of the cell without being thrown out as an alien invader. But they wanted to know how the proviruses managed to "take a seat around the table." Researchers began to look for another enzyme, one capable of inserting the proviruses into the host genome. It didn't take long for scientists to discover a second unique and important enzyme carried in the rucksacks of retroviruses: *integrase.* Retroviruses use integrase to insert proviruses (the DNA copies of their RNA genes) into the host genome. Reverse transcriptase and integrase are criminal partners that disguise themselves and break into the nucleus where they hide out in the host genome. Once they've found their hiding places, the proviruses slowly create havoc in the cell—cancer!

Using the electron microscope and their discoveries of reverse transcriptase and integrase, scientists began looking more closely at what happened once the proviruses took up their positions in the host genome. Because scientists had decades of experience studying MMTV, they looked in mouse breast cancer cells for clues. One of the first things they found was that whenever a cell that was infected with MMTV divided—as it did during pregnancy and lactation—huge quantities of new tumor viruses were also made that were passed down to all the daughter cells. With every division and multiplication of cells, more tumor viruses were made. Under the growth-stimulating effects of female sex hormones, every cell in the breast was soon swarming with reverse transcriptase and MMTV genes. With each menstrual cycle and with every pregnancy over the lifetime of the animal, the genetic "footprint" of MMTV inside the breast enlarged enormously.

Of course, Lathrop and Loeb had found sixty years before the discovery of reverse transcriptase that breast cancer in mice did not appear before puberty, that the risk for breast cancer increased with every pregnancy, and that the overall risk increased over the lifetime of the animal. They reported that the risk for breast cancer dropped dramatically if the ovaries were removed at an early age, thus cutting off the growth-stimulating effects of estrogen and progesterone. Lathrop, Loeb, Bittner, and others understood that female sex hormones played an important role in causing breast cancer in mice, but it wasn't until MMTV was seen with the electron microscope in the 1950s, and was then seen to multiply whenever when breast cells multiplied, that the link between MMTV, female sex hormones, and the risk for breast cancer began to make sense. More viruses inside the cell imparted a greater risk for cancer, but how? The fact that the host genome was packed with MMTV didn't explain causation—the precise mechanism by which the presence of the provirus (the DNA copy of the RNA genes) inside the genome transformed a healthy cell into a malignant one. What was the next step, the cancer-causing step?

The question of the exact mechanism by which MMTV transforms a healthy cell into a malignant cell baffled scientists for years. They could see swarms of MMTV viruses swimming inside breast cells and they could put their fingers on the proviral genes inside the host genome. But they continued to be baffled about how this crowd of retroviruses transformed a healthy cell into a tumor. Then, in 1991, Roel Nusse, PhD, of Stanford University, discovered precisely what MMTV proviruses did inside the host genome that set healthy breast cells on a death spiral to malignancy. Dr. Nusse discovered that when MMTV genes are inserted into the host genome (via integrase), they occasionally land in the vicinity of a gene that, if activated, can set the cell on a course to cancer. These genes are called *proto-oncogenes*, and they are precursors to oncogenes (the genes inside the cell that have the potential to turn a healthy cell into a malignant one). Proto-oncogenes can lie dormant and remain innocuous unless they are disturbed. Then they become oncogenes, which are frankly dangerous. Nusse discovered that when MMTV genes land in the vicinity of a proto-oncogene, the proto-oncogene is activated and becomes a full-fledged oncogene. Once the switch is flipped, the cell grows wildly and abnormally. Eventually it becomes a cancer cell. The cancer cell grows wildly and this creates a tumor.

Here's the recipe for breast cancer: MMTV + reverse transcriptase + integrase + a chance encounter with a proto-oncogene (enhanced by the stimulating effects of estrogen and progesterone) = malignant transformation and a tumor.

When Nusse discovered that the mere presence of MMTV genes in the vicinity of a proto-oncogene was sufficient to activate it into a full-fledged oncogene, he solved the riddle of how MMTV causes breast cancer in mice, cats, dogs, monkeys, and maybe humans, too. The process by which retroviruses, like MMTV, cause cancer by inserting their genes into the host genome and thereby activating proto-oncogenes is referred to as *insertional mutagenesis*. Once Nusse discovered the mechanism of insertional mutagenesis, other puzzle pieces began dropping into place and a better picture of how MMTV causes breast cancer in animals began to emerge. Scientists found that there is a region along the string of MMTV genes that *directly* binds to estrogen. By binding estrogen directly, MMTV uses estrogen like a fuel. So, not only does the cell under the influence of estrogen produce more viruses every time it replicates, MMTV binds estrogen directly and uses it like rocket fuel to make more copies of itself, as well.

Estrogen and progesterone facilitate the cancer-causing effects of MMTV in two ways: (1) by increasing the rate at which the cell divides, which increases the number of viruses made and passed down to daughter cells, and (2) by binding to the virus directly, which causes even more viruses to be made inside the cell. Female sex hormones are powerful amplifiers of MMTV inside the cell and are, thus, powerful promoters of breast cancer in animals infected with this virus.

Dr. Harold Varmus was studying MMTV at UCLA when Nusse discovered insertional mutagenesis. Varmus then began looking for what happened next; that is, the exact mechanism by which MMTV triggered breast cancer once its provirus was inserted into the host genome. What he found was not exactly the answer to this question; instead, he discovered the first oncogene. Varmus found that MMTV proviruses were especially adept at activating certain genes that then turned the cell to cancer. He recognized that these genes (the oncogenes) functioned like cancer switches. Varmus was not able to find evidence of MMTV in human breast cancer, but he was able to find the first oncogene, and he was more than happy with that.

In 1991, Robert Callahan, PhD, a scientist at the National Cancer Institute who had been studying MMTV for years, began a series of experiments to try to understand why some strains of mice had higher rates of breast cancer than others. Of course, this was something that Lathrop had observed nearly a hundred years before. Recall that Lathrop noticed that some strains of mice seemed to get breast cancer while other strains did not. When she learned that the Harvard researcher Clarence Cook Little was interested in using mice to study cancer, she handed him the mice on her farm that got breast cancer, and he used them to create inbred strains in which all of the animals eventually got breast cancer.

Little knew that only certain strains of mice carried genes that increased their risk for breast cancer. When he discovered that there was something else beside the genes that played a role in causing cancer, he asked his scientific team to try to identify the missing ingredient.

Bittner discovered the milk agent, which was later proved to be a virus—MMTV. Bittner recognized that some strains of mice possessed genes that protected them from the cancer-causing effects of MMTV. He was the one who came up with the original recipe for breast cancer in mice: the right genes, the breast cancer virus, and the presence of female sex hormones.

Forty years later, Callahan wanted to understand more about the genes that facilitated the action of MMTV and those that got in its way. Callahan's experiments confirmed that certain strains of mice were genetically predisposed to the cancer-causing effects of MMTV, while other strains carried genes that protected them. His experiments were focused more on understanding the genetic predisposition of MMTV-associated breast cancer in mice and less on making a link between MMTV and human breast cancer. But scientists in other countries who remained undaunted, even when the National Cancer Institute abandoned its Special Virus Cancer Program, pressed full steam ahead to investigate MMTV's role in causing breast cancer in women.

In 1988, Dr. Gary Franklin, MD (Department of Pathology, Royal Victoria Infirmary, Newcastle upon Tyne, England), looked for and found evidence of MMTV genes in human breast cancer specimens, and then he cloned those genes. Franklin published his discovery in a peer-reviewed journal, saying that he had isolated "sequences related to those of the mouse mammary tumor virus . . . from human DNA."[1] This was the first official scientific paper reporting evidence of the mouse mammary tumor virus inside the human genome. Franklin made it clear that he was the first to nail this discovery, declaring that he had found MMTV-related "sequences that have not been described previously" in human cells.[2] He said that he had removed MMTV genes from "five human breast cancer cell lines, from placenta, and from two lines derived from other malignancies."[3] This, of course, meant that the mouse breast cancer virus, or its near equivalent, was capable of infecting human cells! Franklin studied the MMTV genes that he had found inside human breast cancer cells and found that, just as in mice, they were under the influence of estrogen and progesterone.

In 1994, a scientific meeting called "The Challenge of Breast Cancer" was held in Brugge, Netherlands. It was focused entirely on the role of MMTV in human breast cancer. The *Lancet*, a highly respected British medical journal, sponsored the meeting. The meeting was called because "existing attempts to explain the behavior of breast cancer and thereby guide therapy are found wanting in several respects."[4] Scientists running the meeting zeroed in on a recently published paper which broached the subject, still taboo in the United States, of the possibility of retroviral-induced breast cancer in women. Leapfrogging over what was still very preliminary evidence of a role for MMTV in human breast cancer, the men who ran the meeting posed the following question: What antiviral agents (might) prove to be the way forward in treating breast cancer in women? They concluded by saying they hoped for "greater openness . . . in the way in which the collective scientific consciousness is prepared to adapt and to take on broad radically new approaches to prevention."[5]

Chapter 15

Evans, Mueller, and Stewart

You cannot prove that a virus causes cancer in humans by conducting a randomized trial in which you expose half the participants to the tumor virus and give the other half a sugar pill. If you want to demonstrate that a virus causes cancer in humans, you must find an indirect way of providing the proof. But how? What indirect evidence would constitute sufficient proof that a virus causes cancer in humans?

In 1990, Alfred Evans Jr., PhD, of Yale University, and Nancy Mueller, PhD, of Harvard University, established the criteria that would serve as indirect proof that a virus causes cancer in humans. The Evans and Mueller guidelines (now considered the gold standard) fall into two broad categories: epidemiologic and biologic. Let's consider each category separately and then review each criterion in each category separately.

The first category consists of epidemiologic data. Epidemiology is the branch of medicine that deals with the incidence, risk factors, and distribution of a disease in a population. Epidemiology is considered the cornerstone of public health as it reveals the patterns of disease and how they vary across a population. For example, the link between cigarette smoking and lung cancer began with an epidemiologic observation during the 1940s: the incidence of lung cancer suddenly spiked to levels never seen before. Why? Well, epidemiologists took a closer look and found that the spike in the incidence of lung cancer occurred about a decade after the tobacco industry began an aggressive marketing campaign to sell cigarettes to as many people as possible. Cigarette manufacturers put them into GI rations for soldiers headed into World War II and told everyone that smoking was good for your health. Not satisfied with half the population, the tobacco industry marketed directly to women. "Reach for a Lucky instead of a sweet" was one of the most famous advertising slogans Albert Lasker ever wrote.

When the incidence of lung cancer began to rise with the rise in cigarette smoking, epidemiologists hypothesized that there was a link between the two, and scientists began to look for a biologic cause and effect. Pathologist Oscar Auerbach, MD, conducted a series of experiments in dogs in which he hooked them up to respirators that forced cigarette smoke into their lungs. When he examined the tissue under the microscope, he was able to document the slow but inexorable transformation of cells from healthy to cancerous. Thus, it is often epidemiologic observations that provide the first clues about the possible relationship between a risk factor and a disease.

The Evans and Mueller guidelines contain four epidemiologic criteria. The first is the geographic distribution of a viral infection corresponds with that of the tumor it is thought to cause. In the simplest terms, this means that the virus is found where the people with the tumor are found. For example, if, in a given population, 100 people in every 1,000 develop cervical cancer, and there are 100 people in every 1,000 who show evidence of having been infected with the human papillomavirus, HPV, then one might hypothesize there is a relationship between HPV and cervical cancer. In fact, Harald zur Hausen found that this was the case, and this epidemiologic correlation was one of the criteria that were used to establish causation and final proof that HPV causes cervical cancer in women.

The second Evans and Mueller epidemiologic criterion is viral markers are higher in those with the disease than in those without the disease.

As an example, there are more antibodies to the HPV in patients with cervical cancer than in patients without cervical cancer. This criterion, if fulfilled, has important clinical consequences because it means that once all the proof is in and you're sure that, say, HPV causes cervical cancer, then you can screen the population to see who is infected with HPV and follow them more closely for the development of cervical cancer.

The third epidemiologic criterion is a really interesting one, and fun to investigate: evidence of infection (in other words, the presence of viral antibodies) precedes development of the cancer, and the viral antibodies are higher in patients with the cancer than in healthy patients. Let's use HPV as an example. Women only develop cervical cancer *after* they become infected with HPV, not before. And women with cervical cancer have much higher levels of antibodies to HPV than women who do not have cervical cancer.

The fourth epidemiologic criterion is really an endpoint as well as a box that needs to be checked on the way to proof: the incidence of the cancer in question is reduced when viral infection is prevented. Let's return again to HPV as an example. When women are vaccinated against HPV and, thus, are never infected with the virus, they do not get cervical cancer. That is, the vaccine that prevents infection also prevents cancer. This is exactly what has happened in women who have received the vaccine against HPV: the incidence of cervical cancer has dropped dramatically. Just recently it has been discovered that HPV causes other cancers too: 50–65 percent of all head and neck cancers. So, vaccines like Gardasil that prevent HPV infection have the potential to prevent 100 percent of cervical cancer and more than half of all head and neck cancers in the world.

Now we shall consider the three Evans and Mueller biologic criteria for proving that a virus causes cancer in humans. The first of these is the virus transforms living cells. The virus must infect a human cell, and it must transform the cell in such a way that it is abnormal. One of the first things zur Hausen did in establishing proof that HPV causes cervical cancer was to show that HPV was capable of infecting human cervical cells and that once they were infected, the cells were transformed. This was exactly what other scientists discovered when they found that HPV caused head and neck cancers, too.

The next biologic criterion is the viral genes are present only in the cancer cells and not in normal cells. This was found to be true for HPV and cervical cancer.

The last biologic criterion is the virus causes cancer in experimental animals. Because it is unethical to conduct an experiment to see if the virus causes cancer in humans, animals are used as a surrogate model. This also allows scientists to perform in-depth studies of how the virus causes cancer in a variety of animals and provides a wealth of information that would otherwise be unobtainable in humans.

If scientists are to prove that MMTV, or its human equivalent, causes breast cancer in women then they must fulfill all seven of the Evans and Mueller criteria. However, as was the case with HPV and cervical cancer, the proof is nearly obtained when six of the criteria have been met: the criterion that involves demonstrating that prevention of infection lowers the risk for cancer is really the crowning achievement for all the work that has come before. The development of the preventive vaccine, Gardasil, was the last criterion fulfilled in proving that HPV causes cervical cancer; but the proof was really in before Merck agreed to manufacture the vaccine. Now, let's take a look and see where we are in proving that MMTV causes breast cancer in women.

In 1999, Dr. T. H. M. Stewart of the University of Ottawa published a paper that provided epidemiologic evidence of a relationship between MMTV and human breast cancer. He began by pointing out what everyone had known for a very long time—that the incidence of breast cancer varied considerably in different regions around the world. For instance, breast cancer is much higher in the United States than in Japan. While many theories had been put forward to explain the geographic variation in the incidence of breast cancer (most of them having to do with differences in diet or childbearing), no satisfactory answer had emerged. Stewart decided to take a closer look at where around the world various strains of mice lived and compare that to where the women with breast cancer lived. He found that "the highest incidence of human breast cancer worldwide occurs in lands where *Mus domesticus* [the species of mice that carries MMTV] is the resident native or introduced species of house mice."[1] In countries with a high infestation of mice with MMTV, the incidence of breast cancer in women was also high. In countries with a low infestation of mice with MMTV, the incidence of breast cancer was also low. In the areas of the world with the highest infestation of mice with MMTV—the United States and Western Europe—the incidence of breast cancer is also high. Countries with a low infestation of MMTV mice, such as Japan and some regions of China, have low rates of breast cancer. Differences in diet and in patterns of childbearing may also play a role (as Bittner suggested years ago), but MMTV may play an important role too, as Stewart's paper seems to suggest.

If there is a link between MMTV and human breast cancer, then direct contact of infected mice with women—as might happen in crowded and poor communities—or contamination of the food supply (via mouse droppings in grain stores, on farms, in grocery stores and markets) are potential modes of transmission of the virus from mice to women.

Chapter 16

Pogo, Holland, and Etkind

As I mentioned in the previous chapter, one of the criteria for proving that a virus causes cancer in humans is to demonstrate that the genes of the virus are present in tumor cells but not in normal cells. While Dr. Gary Franklin was the first to find evidence of the mouse mammary tumor virus in human breast cancer specimens, Beatriz Pogo, MD, DSc, was the first to show that the virus was present only in the breast cancer cells and not in the normal tissue surrounding the tumor. Dr. Pogo came to the United States from Argentina in the 1970s. She had a degree in medicine, was interested in tumor viruses, and found a research position at Rockefeller University while her husband completed his degree in biochemistry. They were hoping to return to Argentina after his studies were complete, but revolution toppled their plans for going home, so they settled in New York City and continued their careers in Manhattan. Pogo remained at Rockefeller University where she became particularly interested in mouse mammary tumor virus (MMTV) and its potential role in human breast cancer.

Pogo relocated to Mount Sinai School of Medicine, where she continued her investigations of MMTV and eventually became a professor of both medicine and virology. In 1995, Pogo published the results of her work with MMTV and human breast cancer in the journal *Cancer*

Research. She began by saying, "MMTV has been related to human breast cancer in previous studies."[1] Then she addressed one of the criticisms that were offered against such claims by admitting that some scientists believed that the evidence for MMTV in human breast cancer was nothing more than "junk" DNA—old bits and pieces of ancient retroviruses that, over millions of years, had found their way into the human genome. Some scientists claimed that "junk DNA" (also referred to as HERVS, for human endogenous retroviral sequences) was a red herring, not the genuine article (in other words, MMTV). In an effort to address the charge that Franklin's evidence was merely junk—or HERVS—Pogo discovered a portion of MMTV that was not found in any other virus. The portion of MMTV that Pogo found was completely unique to MMTV, and could not be mistaken for any other virus known to man. Nor was this segment of MMTV found anywhere in the catalog of normal human genes.

Having discovered a unique portion of MMTV that could be used as a kind of fingerprint, Pogo began looking for it in human tissues. And then she found it. She found it in 7 percent of benign fibroadenomata of the breast. She found it in 2 percent of normal breast tissue taken from women who had undergone breast reduction. More importantly, she found it in 39 percent of human breast cancer specimens. But she did *not find it in the normal tissue surrounding the tumor*. Pogo then compared the MMTV that was found in mice with the MMTV that she had discovered in 39 percent of human breast cancer specimens: the human version and the mouse version were 95–99 percent identical. She concluded that the fingerprint of MMTV that she had found in human breast cancer cells corroborated similar findings of other scientists and said that MMTV "may play a role in the etiology of a large proportion of human breast cancer."[2]

Pogo's colleague at Mount Sinai was, and continues to be, Dr. James Holland. You may recall that Holland was director of cancer research at Roswell Park Cancer Institute in 1971 when members of the Rogers

Committee paid a visit there during the run up to passage of the National Cancer Act. In fact, Holland testified before the committee about the importance of research on tumor viruses, especially the mouse mammary tumor virus. Holland received his degree in medicine from Columbia University Medical School in 1947, did a residency in internal medicine, and then completed a fellowship in medical oncology. On July 1, 1953, he arrived at the National Cancer Institute on the opening day of its Clinical Center to spend a year there as a research fellow. He then transferred to Roswell Park Cancer Institute and eventually became its director of cancer research.

Holland's early work was with childhood leukemia. In fact, he pioneered the use of combination chemotherapy—the simultaneous use of several drugs to overcome resistance and improve response rates and survival—and then headed up a national cooperative of clinical investigators to evaluate a variety of chemotherapy regimens for treating patients with leukemia. Holland also developed an interest in tumor viruses, particularly MMTV. In 1981, Holland was recruited to Mount Sinai School of Medicine to become the director of its cancer center. He is the editor of one of the premier textbooks in medical oncology, *Holland-Frei Cancer Medicine*, which is now in its ninth edition. Holland is a member of the Expert Advisory Panel for Cancer at the World Health Organization, past president of the American Association for Cancer Research, and past president of the American Society of Clinical Oncology. Over the course of his stellar career, Holland has continued to investigate the relationship between MMTV and human breast cancer and was a coauthor of the paper Pogo published in 1995 in which they reported finding evidence of MMTV in 39 percent of human breast cancer specimens, in 7 percent of benign breast tumors, and in 2 percent of women who had undergone breast reduction surgery.

I was introduced to Dr. Holland's work in February 2007 when my colleague, Ken Blank, MD, gave me a handout from the San Antonio

Breast Cancer Symposium that was held in December 2006. The San Antonio Breast Cancer Symposium is one of the most popular and densely packed scientific venues for the discussion of breast cancer research in the world. Holland presented his research on MMTV to thousands of participants, saying that he and Pogo had actually found a human mammary tumor virus. Why did he refer to it as a *human* mammary tumor virus, or HMTV? Because its genetic signature was unique and distinct from MMTV, and because it had been found in human tissue. Holland also reported that HMTV was capable of infecting normal cells of the breast, as well as a variety of normal human blood cells. HMTV was able to replicate once it had entered and infected a normal human cell, and newly made viruses were then released to infect neighboring cells.

As Pogo had emphasized in her 1995 paper, Holland made a point of saying that the HMTV he had identified was neither junk DNA nor HERVS, but a distinct variant of MMTV that was unique to humans: HMTV. He was emphatic that HMTV is not inherited, as HERVs are and have been ever since they entered the genome millions of years ago and were then passed down the line from one generation to the next. Because HMTV was not found in normal tissue surrounding the tumor, Holland concluded that it was an acquired *infection*. Upon further analysis, he discovered that HMTV was 95–99 percent similar to MMTV. "We thought that if human breast cancers were due to a virus, it would be kissing cousin to the mammary tumor virus in mice," he said.[3]

Note: Scientists had previously identified a mammary tumor virus in cats, rats, and monkeys and had given each of these their own designation; for instance, the virus that causes breast cancer in cats is called feline mammary tumor virus, or FMTV.

Holland reiterated that the segment of HMTV (the "birthmark") that he'd found in the genome of human cells was unique to the breast cancer virus, not found in normal human DNA, nor was it to be found in any other living cell on earth. He told the audience that other scientists had done similar work and that, taken together, the evidence suggested that HMTV was a disease acquired from mice, in the same way influenza or HIV is acquired from other animals. He referenced the work of Stewart, explaining that areas in the world with the highest incidence of mice that are known to carry MMTV are also areas of the world where there is the highest incidence of human breast cancer. Newer data, he said, indicated that this same correlation held true in parts of Africa and Asia. Given his expertise, years of research, and stature in the academic community, Holland was able to suggest with confidence, "Human mammary tumor virus is linked to a large portion of human breast cancer."[4] Holland cautioned that final proof was still outstanding and required at least one additional long-range study. But the point was made: HMTV looked nearly identical to MMTV and behaved like it, too.

One of the early discoveries about the effects of MMTV in mice was that, in males, the virus does not produce breast cancer. Males lack female sex hormones, which is why they don't develop the breast tissue that they are born with. When male mice are infected with MMTV they develop other tumors: lung and lymphoma. Female mice get these tumors too, but the preponderance of their malignancies occurs in the breast as a result of stimulation by estrogen and progesterone.

There are some interesting correlations in human patterns of disease. It's long been known that women with breast cancer have a higher incidence of lymphoma. Dr. Polly Etkind, a virologist who was working at the Sloan Kettering Cancer Institute when Pogo published her work about MMTV in 1995, decided to repeat Pogo's experiments in her laboratory at Sloan Kettering. And when she looked, Etkind found essentially the same thing, and more. Etkind found evidence of MMTV in 37 percent of the human breast cancer specimens she examined, finding no evidence of the virus in normal surrounding tissue. And then, knowing that MMTV causes both breast cancer and lymphoma in mice, Etkind examined specimens taken from women who had been treated for breast cancer and who later developed lymphoma. In 57 percent of cases, both tumors showed evidence of MMTV. Etkind then examined the tissue of patients who had only been diagnosed with lymphoma: three out of nineteen patients had evidence of MMTV in their tumors.[5] Etkind published a paper corroborating Pogo's data and adding new evidence of MMTV's possible role in causing human lymphoma. She believed that the breast cancer virus—of which there are several variants—resides predominately in the mouse as MMTV, and by some means yet to be discovered, moves into humans as HMTV. Direct exposure to mice, or transmission of the virus by another mechanism—our food supply or our pets—are still mysteries waiting to be solved.

Chapter 17

Lawson, Ford, Garry, and Levine

James Lawson, MD, is a professor of public health at the University of New South Wales, Australia. In 2001, he published a commentary in the journal *Breast Cancer Research*, "From Bittner to Barr: A Viral, Diet and Hormone Breast Cancer Aetiology Hypothesis."[1] Lawson began by saying that scientists had been suggesting for years that the human equivalent of the mouse mammary tumor virus (MMTV), in conjunction with cofactors like diet, hormones, and a genetic susceptibility, played a role in human breast cancer. He encouraged the scientific community to pursue this possibility without further delay because "all of these hypotheses are testable."[2]

One of Lawson's colleagues, Caroline Ford, PhD, began looking for evidence of human mammary tumor virus (HMTV) in women diagnosed with breast cancer in Sydney, Australia. In 2004, she published the results of her investigations in the journal *Cancer Research*.[3] Not only did Ford discover evidence of HMTV in patients with breast cancer, she also found that increasing levels of the virus were associated with increasing risk for, and aggressiveness of, the disease—in both men and women. The title of Ford's paper summarizes her findings: "Progression from Normal Breast Pathology to Breast Cancer is Associated with Increasing Prevalence of Mouse Mammary Tumor Virus-Like Sequences in Men and Women."

Ford began her study of MMTV in humans by first replicating the work of Holland and Pogo. She examined tissue specimens taken from women who had been treated for breast cancer at the Prince of Wales Hospital in Sydney and found that approximately 40 percent had evidence of MMTV in their tumors. These results are nearly identical to those Pogo and Etkind had reported earlier. In Ford's study, only approximately 2 percent of breast tissue removed from women with benign biopsies showed evidence of MMTV. Again, these results were identical to the results that Holland and Pogo had reported earlier. But then Ford expanded her investigation to see if there was a "dose-response" relationship between MMTV and breast disease; that is, she looked to see if there was a relationship between increasing levels of the virus and the risk for and aggressiveness of breast cancer. "Screening of a larger and more diverse cohort of female breast cancer samples has now shown a correlation of MMTV-like sequences with the severity (grade) of breast cancer," she reported. "A significant gradient of MMTV positivity was observed with increasing severity of cancer."

Then for the first time, Ford reported finding the virus in men with breast cancer. In fact, her results were even more astonishing in men than women: 62 percent of men with breast cancer showed evidence of the virus; 19 percent of men with gynecomastia (benign breast enlargement) showed evidence of the virus. Ford concluded, "These results support the association of MMTV-like sequences with development of breast tumors in men and women and suggest association of MMTV with increasing severity of cancer."[4]

Ford's discovery addresses one of the Evans and Mueller criteria for proving that a virus causes cancer in humans. Specifically, investigators must show that there is a higher prevalence of MMTV in women with breast cancer than in healthy women. Holland, Pogo, and Etkind had shown this for women in the United States, and now Ford reported finding the same thing in women in Australia. A corollary to this positive relationship, referred to as an "increasing viral load and subsequent increased risk for cancer," is to also demonstrate that increasing levels of the virus (in other words, the viral load) are associated with increasing aggressiveness of the disease. When a positive correlation exists between increasing viral load and increasing risk for and aggressiveness of breast cancer it is called a *biologic gradient.*

This dose-response, positive-correlation, biologic gradient can be understood by taking a simple metaphor from the baker's repertoire. You've made, say, a birthday cake, and now you want to frost it with icing. You decide you want the icing to be hot pink in color. You make up a bowl of white butter cream frosting (this is making me hungry), and you add one drop of red dye. The frosting turns light pink. One more drop and the color becomes "race for the cure" pink. One more drop and the frosting turns hot, hot, hot pink: you don't dare add another drop because this will take your masterpiece into the red zone. In a similar fashion, adding increasing amounts of MMTV to normal breast tissue (under the influence of estrogen, age, carcinogens, alcohol, smoking, and so on) yields the same results as our test kitchen case.

A little virus is associated with benign breast tissue. The addition of more viruses leads to a premalignant tumor (ductal carcinoma in situ, or DCIS; Ford reported this, too). A bit more virus gives you a low-grade invasive breast cancer. Add more viruses to the mix and you get a very aggressive breast cancer, until you get to the point where the "frosting" is completely red! There's enough virus to produce the deadliest breast cancer known to man: inflammatory breast cancer. In April 2015, Holland presented his most recent data and at the annual meeting of the American Association of Cancer Research in which he reported finding HMTV in 94 percent of specimens taken from women with Stage 4 breast cancer. (He calls it HMTV, whereas Ford still calls it MMTV.) Which is to say, as the disease progresses and finally gets out of hand entirely, the footprint of the virus in the body becomes extremely large.

The dose-response, biologic gradient that Ford demonstrated in her study of Australian women was repeated in a study of women from Vietnam. Again, she reported that increasing levels of MMTV increased the risk for and the aggressiveness of breast cancer in women. At the end of her paper, Ford said, "These data, taken together, add to the growing number of studies implicating a MMTV-like virus in human breast cancer, although a clear causal relationship of MMTV to breast cancer remains to be established."[5] Yes, it does.

Another important Evans and Mueller criterion for proving that a virus causes cancer in humans is to show that the virus can produce tumors in experimental animals. Of course, we've known for about one hundred years that the virus causes breast cancer in mice. But its role in causing breast cancer in cats is particularly important when considering how the mouse virus makes its way into the human population. Robert F. Garry, PhD, is a professor of microbiology and immunology at Tulane University. Most recently he has been involved in helping researchers at the Harvard School of Public Health obtain blood samples from patients infected with Ebola in West Africa; but he has also spent years investigating retroviruses like HIV and MMTV. In 2005, Garry published a paper on the existence of the breast cancer virus in cats, what he called the "feline mammary tumor virus, FMTV."[6] Garry reported that FMTV was nearly identical to MMTV in the same way that HMTV is nearly identical to MMTV. In his paper, Garry proposed that cats acquired the virus from infected mice. He hypothesized that cats pass the virus to their offspring via breast milk, just as mice do. He also proposed that cats infected with FMTV might be a source of transmission of the breast cancer virus to humans, especially when their pets are pregnant or breastfeeding. In his paper, Garry referred to the work of other scientists who had shown that MMTV was capable of infecting normal cat breast cells, demonstrating that the virus had the ability to "jump species" from mice to cats. Garry used FMTV that he obtained from cats to successfully infect normal mouse breast cells. Indeed, the virus could "jump" from cats to mice and from mice to cats. This helped to confirm that the virus was transmissible among different species, something that has also been demonstrated when MMTV is introduced into normal human cells.

There was other news to report, as well. Garry discovered that once the breast cancer virus had jumped from mice to cats, it acquired the ability to move into a larger range of species. This observation suggested that the ability to jump species meant that the virus was more robust, as if this exercise had made it stronger and more threatening to other animals. Garry hypothesized that, as a result, FMTV might actually be more infectious in humans than MMTV. Garry envisioned the day when pets would be screened for the mammary tumor virus and/or given a vaccine to prevent the contagion from infecting their owners.

Garry continued his investigations with FMTV, reporting his research at the Eleventh International Congress of Virology in Sydney, Australia, in 2010. He suggested that HMTV might be one of the factors that triggered breast cancer in women, and that, as in mice, it may be an inherited form of the disease (like a BRCA mutation) as well as a disease that is passed horizontally (from one infected animal to another). He said, "If a definitive link to this retrovirus is established, HMTV may become a target for a vaccine to prevent breast cancer and a target for new treatments for breast cancer."[7] Garry reported that he had looked for evidence of MMTV in the tumors and organs of women with breast cancer and had found it, just like Pogo did. Like Holland, Pogo, Etkind, and Ford, Garry reported that he found HMTV in a small portion of healthy women who did not have breast cancer. Garry suggested, as Bittner had decades before, that other factors, such as an individual's genes, likely contributed to the potential of the virus to cause breast cancer in women.

Another Evans and Mueller criterion for proving that a virus causes cancer in humans is to show that the viral genes are present in tumor cells, but not in normal cells. Franklin, Holland, Pogo, and Garry had shown that MMTV genes were present in human breast tumors, but were absent from normal surrounding breast tissue. In 2010, Dr. Paul H. Levine, MD, professor of epidemiology at the George Washington University School of Medicine, reported finding the virus—he referred to it as HMTV—in the majority of women with inflammatory breast cancer. Inflammatory breast cancer (IBC) is the most aggressive form of breast cancer known to humanity. It's also the most sinister, for its symptoms look very similar to those associated with a breast infection. All of a sudden, one day, a small portion of skin on the breast becomes slightly red. There might even be a small red spot as if a bug had bitten the patient. The next day, the redness will be more pronounced. The breast might feel warm to the touch, but not always. In a matter of days, the breast will swell. The skin will become edematous (filled with fluid) and will develop the appearance of an orange peel, called *peau d'orange*.

Typically, the patient goes right to the doctor who diagnoses a case of breast infection (mastitis) and prescribes an antibiotic. It doesn't help. The lymph nodes under the arm begin to swell. The breast begins to look a whole lot worse. The patient returns to the doctor, and the doctor switches antibiotics to see if something "stronger" might work better. Nothing works at all, and the patient continues to deteriorate.

Occasionally the redness creeps across the chest to the other breast. Lymph nodes under the opposite arm begin to swell.

Weeks may go by, sometimes months. The patient, and then the doctor, panic, and finally the patient is sent to a breast specialist. A biopsy is done. It's cancer of the worst kind. It's inflammatory breast cancer. The prognosis is dismal. Chemotherapy is begun immediately. Surgery follows. Lumpectomy is not an option: the breast must be removed. Often the other breast must be removed as well if the disease has progressed that far or threatens to do so.

The chest wall is irradiated following surgery. More chemo may be given. Reconstruction is never done at the time of surgery for the risk of local recurrence is much too high.

Women who have a complete response to all this treatment are still at a very high risk for recurrence, usually sooner rather than later. Only 25–50 percent of patients with IBC survive five years, compared to 90–95 percent of women who are diagnosed with other types of breast cancer. IBC represents less than 10 percent of cases of breast cancer in the United States, but in parts of Africa, it's the most common form of the disease.

Dr. Levine, working with Dr. Pogo, began an investigation of IBC in Tunisia. He found that approximately 70 percent of women with IBC had evidence of HMTV.[8] Then he took a look at IBC in the United States and in 2010 reported finding HMTV in 71 percent of women with IBC here.[9]

Chapter 18

Pascrell, Niederhuber, Callahan, Rouault, and Arbus

In September 2009, I met with my congressman, Bill Pascrell, in Washington, DC, to tell him about my investigation concerning the history of research on the breast cancer virus and share with him my concern that this subject was getting short shrift by the major breast cancer foundations and the National Institutes of Health. I asked for his help, and on October 29, 2009, Congressman Pascrell wrote to John E. Niederhuber, MD, then director of the National Cancer Institute, asking for the following information. "How much money has been spent by the NCI over the past five years researching the link between a virus and human breast cancer, including amounts provided to outside groups or organizations? Are there plans to allocate funds from the FY2010 or future budgets for research into this viral link? What specific research studies have been funded by the NCI in-house or at other institutions regarding viral links to human breast cancer?" These were three very simple, straightforward questions.

Dr. Niederhuber is a public servant. Until 1971, the director of the National Cancer Institute was a member of the Public Health Service who had risen in the ranks. After the passage of the National Cancer Act, the director of the National Cancer Institute became a political appointee. Nevertheless, his salary is paid for by taxpayers. Now, the director of the

National Cancer Institute may not feel the need to answer every question put to him by curious citizens, but it's fair to expect that he would answer direct questions put to him by a member of Congress, particularly one as senior as Congressman Pascrell. Three simple and straightforward questions ought to have produced three simple and straightforward answers.

On December 11, 2009, Dr. Niederhuber sent a two-page letter of reply to Congressman Pascrell, in which he failed to fully answer any of three questions asked of him. I include the letter in its entirety here for your amusement:

> Thank you for your letter dated October 29, 2009, describing your interest in research on viral links to breast cancer. I am pleased to share with you some information about how the National Cancer Institute (NCI) is working to understand more about viruses and their roles in human cancer development.
>
> Viruses are now accepted as authentic causative factors for a number of human cancers. Approximately 20 percent of all cancers worldwide are thought to be etiologically associated with viral infections. Several viruses have been definitively linked to specific malignancies in humans; for example, we know that most cervical cancers are caused by Human Papillomaviruses (HPV), and the associations of Hepatitis B and C viruses with human liver cancers are now well established. There are additional examples, as well, and it is reasonable to expect that other associations might be discovered.
>
> It should be noted that other microbes, in addition to viruses, are responsible for a large part of the global burden of infection-associated cancer. For example, the bacterium Helicobacter pylori is the most important risk factor for the development of gastric cancer. NCI is committed to conducting research on relationships between infectious agents and cancer development progression.

Finding definitive links between cancer and viruses presents some interesting challenges. Most of the viruses that have been linked to cancer are very common in human populations; yet it is rare and takes a long time for an infection to develop into cancer, making identification of the causal agent difficult. In addition, even cancers that have proven viral etiology do not develop solely as a result of viral infection—other factors are required.

Recent intriguing findings suggest new cancer associations for some well-characterized viruses as well as previously unknown viruses, and NCI is supporting a range of projects to explore these possible links. Included among them are a wide variety of studies of HPV and possible relationship with oral, lung, and esophageal cancers. A consortium of investigators is evaluating gastric cancer risk associated with Epstein-Barr virus (EBV) infections. NCI-supported research recently demonstrated the presence of a novel retrovirus, Xenotropic Murine Leukemia-related virus (XMRV) in prostate cancer tissue samples. The XMRV research community is currently developing and validating research tools to accurately test for XMRV in tissue samples, and to understand what role, if any, XMRV may have in prostate cancer etiology.

In your letter, you indicated an interest in learning more about Mouse Mammary Tumor Virus (MMTV). Although MMTV provides an excellent animal model for viral cause of breast cancer, research, to date, has failed to identify a human version of the virus. Some, but not all, viruses operate in multiple species. Most are species-specific. Nevertheless, scientists continue to pursue a human equivalent for MMTV. An NCI case-control study of inflammatory breast cancer in North Africa is ongoing and will include testing for a human MMTV in tissue samples. It is thought that if such a virus were to exist, it would most likely play a role in the development of this rarer form of breast cancer that is common in this region.

> NCI has continuously pursued research on viral links to cancer as part of our broad portfolio of research on cancer causation. We estimate that we invested $954,858,711 during fiscal years 2004 through 2008 (the most recent five-year period for which actual expenditure data is available) in experimental or epidemiological studies of viruses as carcinogens or co-carcinogens. Current research solicitations could lead to additional awards in this area in fiscal year 2010, and future years. Investigator-initiated grant applications are accepted on a continuous basis and we rely on a rigorous peer-review system to identify and fund the most promising and highest quality research. NCI will continue to leverage its expertise in microbiology, cancer biology, immunology, and clinical research to advance scientific knowledge of viruses, and other infectious agents, associated with breast cancers, and other types of cancer. I appreciate very much your interest in our work, and hope this information is useful to you. Please do not hesitate to contact me if you have further questions or at any time I may be of assistance.

Not only did Niederhuber dodge the three questions put to him by Congressman Pascrell, he completely negated the many scientific papers that had been published in peer-reviewed journals in the United States and around the world indicating that MMTV does play a role in a significant portion of human breast cancer. I was stumped by Niederhuber's slick stonewalling of Congressman Pascrell, and I vowed to find out what exactly was going on at the NCI with funding of research on the human mammary tumor virus.

I called Niederhuber's office and asked to meet with him. I was told that I could not meet with Niederhuber, but that the NCI would allow me to meet with Dr. Robert Callahan, the scientist who'd been working with MMTV at the NCI for years. Dr. Callahan was cordial and attentive. I asked him why the NCI had shut down research on the question of MMTV and human cancer here in the United States. He said that a conference had been held at the NCI in the 1980s and that, at the time, there were insufficient data to support the research. I mentioned the more recent papers published around the world making the case for HMTV, the human breast cancer virus.

Yes, he was aware of them, but not yet persuaded to lift the drawbridge to funding. I asked him what it would take for the NCI to reconsider its closed-door policy.

Callahan's answer was simple and straightforward: He wanted the *flanking sequences.*

I kept a poker face and pressed him further: "You're telling me that if scientists provide you with the flanking sequences you will open the doors and support this research?"

"Yes," he said.

"Okay," I said.

We parted cordially and I spent the rest of the day speaking to other staff members at the NCI who told me they were very interested in helping if, and when, they could.

Why the poker face? I had no idea what flanking sequences were. I called Etkind from my cellphone as I was leaving the NCI that day. I told her what had happened, and what Callahan was looking for, and that he would open up funding for HMTV in the United States if he was provided with the flanking sequences, and could she please explain to me what they were?

Flanking sequences are the genetic sequences made of DNA that sit on either side of where the virus "lives" in the host genome. Think of flanking sequences as a kind of address. Flanking sequences are where the virus lives among the other genes that comprise the human genome inside the nucleus of the cell. Callahan wasn't satisfied with the DNA evidence of HMTV that Holland, Pogo, and others had reported. He wanted to know exactly where the virus lived. He wanted the address of the virus. He wanted the flanking sequences. Of course, I had no idea how hard flanking sequences are to find, now that I had just learned what they were. But nothing could have prepared me for Etkind's comment after she explained to me what they were: "We have them!"

Meanwhile, scientists elsewhere in the world marched on in their hunt for a human breast cancer virus. Dr. Francoise Rouault, PhD, working at the Research Institute for Virology and Biomedicine at the University of Veterinary Medicine in Vienna, Austria, published a paper in 2007 showing that MMTV could successfully infect and replicate in human breast cells.[1] Rouault also reported that once a single human breast cell was infected with the virus, every other cell in the culture became infected too. Rouault then exposed the human cells to AZT, a drug used to prevent the spread of HIV. AZT blocks reverse transcriptase, the "secret sauce" that retroviruses use to park DNA copies of their RNA genes into the host genome. Rouault found that, as with HIV, AZT effectively prevented MMTV from infecting human cells. Rouault referred to the work of Holland and Pogo, saying that the skepticism about the existence of a human mammary tumor virus was largely a matter of "deep-seated dogma." She said that humans were undoubtedly hosts for MMTV. Rouault concluded, "Despite the widely accepted belief that human cells are not appropriate hosts for MMTV, our data demonstrate the productive infection of human breast cells. This finding might help to explain the presence of MMTV-like sequences in at least a proportion of human breast cancers."[2]

In 2012, I was invited to join the Leadership Council of the Harvard School of Public Health (HSPH). One of its members, Loreen Arbus, agreed to host a meeting of scientists and concerned citizens to discuss research on the human breast cancer virus in her home in New York City. This took place in April 2013, and at Arbus's request, it was organized by the HSPH. Scientists from the HSPH and from Columbia University joined with Dr. Pogo to discuss her research on HMTV with activist women in New York City. Professor Vincent Tuohy of the Cleveland

Clinic, who developed the first preventive breast cancer vaccine in mice in 2010, also joined the discussion.

By 2013, other stakeholders in the breast cancer community had finally come around to appreciating the potential of this research. The Avon Breast Cancer Crusade gave Dr. Pogo a small grant. The Susan G. Komen for the Cure Foundation gave Holland a small sum of money. Dr. Susan Love's foundation did not fund any of this research, but she did go on the record saying that she wouldn't be surprised if a virus caused breast cancer.[3] The National Breast Cancer Coalition had created the Artemis Project in late 2010 with the goal of developing its own preventive breast cancer vaccine (in response to the Cleveland Clinic's announcement that it had developed one in May 2010) and added to its mission the hunt for a human breast cancer virus.

The momentum in favor of investigating the role of HMTV and human breast cancer began building in a serious way seven years ago when I first got started hammering away on the topic, but the real tipping point came, I think, in July 2013, three months after Arbus's summit for the "pink virus" in New York City.

Chapter 19

The National Institutes of Health

On Wednesday, July 10, 2013, the Clinical Center at the National Institutes of Health, the largest research hospital in the United States, celebrated its sixtieth anniversary. The Clinical Center has treated nearly a half million patients and produced some of the most important breakthroughs in cancer research, diagnosis, and treatment. On this particular day two years ago, as of this writing, hundreds of guests were invited to attend a special grand rounds series of lectures on research milestones, emanating from the Clinical Center.

John Gallin, MD, director of the NIH's Clinical Center, gave the opening address for grand rounds that day. He spoke for about forty minutes and then took another fifteen minutes to introduce the first speaker, Dr. James Holland. In introducing Dr. Holland, Gallin remarked, "How lucky we are to have him present his work."[1] Dr. Holland is now ninety years old, but he is still as sharp and vigorous as any young man. He took the podium saying that Gallin's enthusiastic introduction reminded him of an obituary, a premature one he hoped. Holland's topic that day was the "Human Mammary Tumor Virus."[2]

Holland began his lecture by telling the story of his career. He arrived on opening day of the Clinical Center sixty years before with an interest in leukemia. James Watson and Francis Crick had, only three months earlier, discovered the structure of DNA. These were still early days in modern medicine. Sidney Farber had pioneered the use of single-agent chemotherapy—one drug given at regular intervals—in treating children with leukemia. He was one of the first to report encouraging results. Farber showed that median survival for acute childhood leukemia could be extended in about 10 percent of children, sometimes for as long as four months, sometimes longer. That was a huge breakthrough, but nowhere near enough. Holland thought that by adjusting the dosing schedule and using more than one drug to overcome tumor resistance that survival might be prolonged beyond a few months. He was right. He tried a combination of drugs and gave them every few days, and he achieved remarkable results. Over the next several years Holland showed that this new regimen in which he used a combination of drugs would lead to very extended survival rates: 60 percent of patients with acute leukemia were still alive and well ten years later.

In 1954, Roswell Park Cancer Institute offered Holland a 30 percent increase in his salary, and so he left the NCI, but not entirely. He was asked if he would be willing to collaborate with the NCI to begin the first multi-institutional collaboration in cancer research—and the one that has lasted the longest. Holland then began to discuss the topic he was invited to present: the breast cancer virus. "I was fascinated by the observations made by John Bittner in 1936," he said. Bittner had discovered that there was a milk agent, which turned out to be a breast cancer virus passed from infected mothers to their offspring via breast milk. Bittner reported that mice who do not themselves have breast cancer can give the virus to their offspring via infected breast milk, and pass the disease on to subsequent generations in the same manner.

Holland told the audience that after the mouse pup ingested milk infected with the mouse mammary tumor virus, the virus moved into the white blood cells (lymphocytes) that line the digestive tract. The infected lymphocytes then circulated in the bloodstream. When the lymphocytes infected with MMTV arrived via the bloodstream in the breast, they "unloaded" the virus into the breast cells. Later, female mice would develop breast cancer. "Sometimes [the mice] also got lymphoma," Holland said.

Holland then presented Dr. Beatrix Pogo's research. For those in the audience who didn't know her, Holland told them she was a physician and a virologist who had been studying tumor viruses in mice at the Rockefeller University when the scientist she was working for, Charlotte Friend, PhD, died. Holland had been recruited to Mount Sinai University to become the director of its Cancer Center, and he knew of Dr. Friend and of Pogo's work. Because he was interested in MMTV, he invited Pogo to join him at Mount Sinai. Pogo told Holland that she would leave Rockefeller only if she could move from mice to humans in her investigations of the breast cancer virus. He agreed, and they've been working together ever since.

Holland then launched into the meat of his talk. "We found a sequence of the envelope gene in MMTV that was absolutely unique," he said. This segment of DNA "didn't occur in any other virus," and "it didn't occur in the human genome." To drive the point home, he said, "Finding it we knew we would have a 'footprint' of the virus [MMTV]." Holland mentioned that MMTV inserts its genes, via reverse transcriptase, "indiscriminately in several different chromosomes." Which is to say, it takes a seat in the host genome wherever it can find an opening.

After having identified a unique "footprint" of MMTV, Holland and Pogo began looking for the virus in human breast cancer specimens. They searched the archival tissue of the National Cancer Institute Breast Cancer Tissue Registry. Three institutions contribute specimens to this tissue bank: Fox Chase Cancer Center in Philadelphia, the University of Miami in Miami, and Washington University in St. Louis. And they found the breast cancer virus in a significant portion of the breast cancer specimens but not in the normal breast tissue that had been removed around the tumor. Holland made it clear: "Women acquire this virus." He referred to the virus, as the title of his presentation suggested, as the human mammary tumor virus, HMTV. He gave it that name, he said, because "it came from human tissue."

Holland then projected an image of the virus for the audience to see: "It looks just like MMTV," he said. Holland acknowledged the man who first showed an image of the breast cancer virus in human tissues, Dr. Daniel Moore. He had published his photo of HMTV "15–20 years before us." And then Holland put up a slide with the flanking sequences for the virus, the "addresses" where it lives in the human genome—the sites that Dr. Robert Callahan had been looking for—saying, "And we have insertion sites." Holland had found more than sixteen insertion sites. He paused a moment there and then went on. "MMTV and HMTV do not contain oncogenes of their own," he said, confirming what had been known for years. "They [HMTV genes] get randomly inserted" in the host genome, and "they may impact on nearby genes that may be known oncogenes," as Nusse had demonstrated when he discovered insertional mutagenesis as the mechanism by which MMTV leads to breast cancer in the mouse.

As Levine and Ford reported earlier, the viral loads of HMTV in human breast cancer specimens varied according to the type of tumor. Overall, 41 percent of women with breast cancer in the United States showed evidence of HMTV. Holland also discovered that if the patient had a sister or a mother or an aunt with breast cancer—that is, if there was a familial pattern to breast cancer—then there was evidence of HMTV 63 percent of the time. If the patient had been pregnant or was breastfeeding when she was diagnosed with breast cancer, there was evidence of HMTV 64 percent of the time. If the patient had been diagnosed with inflammatory breast cancer (IBC), there was evidence of HMTV 71 percent of the time. Holland explained why there was evidence of an increased viral load in patients with pregnancy-associated breast cancer. As Loeb and Lathrop suggested, and Bittner later hypothesized, "there are hormone-responsive elements present in the genes of both MMTV and HMTV."

Holland then introduced the work of Dr. Paul Levine at George Washington University. Like Holland, Levine was an alumnus of the National Cancer Institute and has since maintained close ties there. Levine had found that women in the United States and in Tunisia had universally high levels of HMTV in their tumor specimens, on average 70–75 percent. Holland mentioned the work of Dr. Thomas Stewart of Ottawa, who had reported on the relationship between the geographic variation in the incidence of breast cancer around the world and the geographic distribution of mice that carried MMTV. In Eastern Europe and Asia, the most common type of mouse (*Mus muscularis*) is a low carrier of MMTV. In Western Europe, the United States, Australia, and western Africa, the most common type of mouse (*Mus domesticus*) is a high carrier of MMTV. The incidence of breast cancer was highest in regions where the mouse that carried high levels of MMTV (*Mus domesticus*) was the most common species. In fact, the incidence of breast cancer in women was three times higher than in regions like Japan, China, Korea, and Thailand, where *Mus muscularis* was the most common species.

Holland then looked to see if Stewart's findings held up under further investigation. He found that in regions of the world where there was a low incidence of breast cancer, there was also a lower incidence of HMTV in the breast tumors. For instance, only 3 percent of the women in Japan had evidence of HMTV in their breast cancer specimens, a country where the incidence of breast cancer is very low. Another country where the incidence of breast cancer is very low is Vietnam; only 4 percent of the women had evidence of HMTV in their breast cancer specimens. In Mexico, Brazil, Argentina, and Australia, where the incidence of breast cancer is relatively high, the incidence of HMTV in breast cancer specimens ranged from 31–38 percent.

To explain the interesting geographic variation in the distribution of different species of mice and its relationship to the incidence of breast cancer worldwide, Holland explained that, historically, West African countries, such as Cameroon, Zaire, Congo, and Tunisia, have been active traders with countries in Western Europe, which provided access to the high-carrier MMTV species of mouse (*Mus domesticus*) that traveled along on trading ships. Because women in West Africa with breast cancer have been found to have evidence of HMTV 60–80 percent of the time, a genetic disposition must play an important role in promoting breast cancer, too, just as Bittner had suggested almost a century ago.

Holland continued: "We had to show the virus was infectious." That is, Holland had to demonstrate the HMTV that he had removed from human breast cancer specimens was capable of infecting normal human cells. And so, he took normal human blood and breast cells (obtainable from laboratories that supply for research) and he made sure that there was no evidence of MMTV or HMTV in their normal genomes: he didn't want "contamination" to confound his results. Then Holland exposed the normal human blood and breast cells to HMTV and looked for evidence of infection. He found that 100 percent of the cells from two types of human blood cells were infected when exposed to HMTV. And he found that half of normal human breast cells became infected when exposed to HMTV. Having proven that HMTV was capable of infecting normal human blood and breast cells, and then burying its genes inside the human genome, Holland checked to see if, as a result of the infection, the cells were transformed. He wanted to know if HMTV infection begins the cascade of changes that ends in cancer.

Holland tested the infected cells for a protein that is associated with malignant transformation of normal cells. That protein is *vimentin.* The production of vimentin is a characteristic of cancer cells. The normal human blood and breast cells that Holland infected with HMTV produced vimentin. They were transformed. They were on their way to becoming cancer. Holland told the audience that the "cells have acquired cancer characteristics as a result of infection with HMTV."

Holland then mentioned the work of Polly Etkind, PhD, the scientist at the Sloan Kettering Cancer Institute who first replicated Pogo's work in the 1990s. Etkind had found that 50 percent of women whose breast tumors were positive for HMTV and that subsequently developed lymphoma had evidence of HMTV. "Exactly like the mouse," Holland said. Holland reported another cancer in which HMTV had recently been found 25 percent of the time: endometrial cancer. He mentioned that patients with biliary cirrhosis, a benign, but deadly, liver disease had evidence of HMTV, too.

Having reviewed a hundred years of accumulating evidence that pointed to the existence of a human breast cancer virus, Holland addressed the elephant in the room: How do women become infected with the breast cancer virus, HMTV? He discussed a case report of a woman who worked as technician in a laboratory that studied MMTV-infected mice. She had been tested on several occasions for evidence of MMTV infection, and on three separate occasions her blood tests had come back negative. Then, one day, her blood test was positive for the virus. Nine months later, she was diagnosed with breast cancer. Several months after that, a lymph node under her arms also tested positive for the virus. Holland said, "The mice of this world contaminate our food chain, up and down the food chain." He said that for a time he had collaborated with the Food and Drug Administration (FDA) to look for evidence of the virus in storehouses where our food supplies are warehoused. The FDA was able to confirm infestation of the warehouses with mice (*Mus domesticus* that carry MMTV), by using ultraviolet light to identify where the mouse had urinated on the food. But the FDA had to cease its investigations because it was short of manpower and had become concerned about the poisons that contaminated imported food. "Too bad," Holland quipped.

Holland then turned his attention to an even bigger elephant no one dared to mention: human breast milk. "We found, as in the mouse, normal women shed HMTV in the cells in their milk." Holland had discovered that 8 percent of healthy mothers had evidence of HMTV in their breast milk. Then he did the math: there are four million births in the United States every year. Half of them are female. He said, "Eight percent of two million women is enough to account for the 40 percent of the 240,000 cases of breast cancer every year . . . *if this were the only means of transmission.*" He reiterated that other scientists had found MMTV in the breast milk of healthy women. Holland pointed out that the likelihood of finding MMTV in human breast milk was higher if there was also a history of breast cancer in the family.

Forty years ago, at least one researcher suggested that women with a family history of breast cancer might want to avoid breastfeeding their children. Bittner had said the same, but in more understated terms. Holland threw down the gauntlet: "We think this is likely the mechanism by which it [HMTV] is transmitted."

Holland concluded, "We haven't proved that HMTV causes breast cancer yet. But of course, we are following that." His son would be next to speak that day. Before he left the stage, Holland said, "If I can't prove that a virus causes breast cancer, maybe he or his colleagues can." The audience gave him a rousing ovation, as the next generation of investigators follows in his wake.

Chapter 20

Seven Billion People

The first recorded case of breast cancer is found in the *Edwin Smith Papyrus*, an Egyptian manuscript that is more than 3,500 years old. Millions of cases of breast cancer have come and gone in the meantime. Some famous cases stand out: Raphael's mistress (as depicted in his painting *La Fornarina*), President John Adams's daughter, First Lady Betty Ford, the first female Supreme Court justice Sandra Day O'Connor, my mother, and, now, a half million women a year. And still we don't know why. But I think we're getting close. One hundred years ago, a retired schoolteacher in Massachusetts, Abbie Lathrop, stumbled upon the first clue and didn't ignore it: mice on her farm developed breast cancer. She turned them over to Clarence Cook Little and Leo Loeb when she found out they were interested.

When John Bittner looked a little closer, he found an infectious agent in the milk and later showed it was a virus. Decades of research by, literally, thousands of scientists now suggest that a virus that is nearly identical to the one that causes breast cancer in mice (MMTV) plays a role in 40–94 percent of human breast cancer. Only one criterion of the Evans and Mueller guidelines remains to be completed before we know for sure. That is, we must enroll a group of women in a study in which the volunteers are followed for several years to determine (via serial blood testing) who becomes infected with the virus, and who does not, and determine whether infection precedes the onset of breast cancer and increases the risk for it. If it does, the last steps will be to make a preventive breast cancer vaccine against HMTV, identify those women whose breast cancer was caused by the virus (so that they can be given antiretroviral therapy, if that is shown to help), and identify those women who are infected with the virus but have not yet developed breast cancer (so that they can be put into a high-risk screening program and given antiretroviral therapy, if that is shown to help.) Sadly, less than $100,000 of the hundreds of millions of dollars that are spent annually for breast cancer research is given to support work on the human mammary tumor virus. That doesn't add up.

Over the past century, the scientists who have worked to prove that a virus causes breast cancer in women have been patient and methodical in making their case. The academic community, ever cautious and slow to change, has been slow to come around to the accumulating evidence. But, in inviting Holland to present a summary of his research on HMTV at the sixtieth-anniversary celebration of the Clinical Center, the National Cancer Institute has, in essence, given the virus its blessing. Holland has said publicly that he hopes to live to see the day the final proof is in. Pogo (who is in her late seventies) told me that she believes, as she gets older, that the virus is keeping her alive.

Is it possible that we've come this far and are wrong? Yes, of course. It is possible to be, say, 70 percent sure that the HMTV causes breast cancer in women only to find out that it does not. It is not likely, given the preponderance of accumulated evidence; but it is possible. How much longer should we wait to know? How many more women must be maimed, or worse, die, before the research is completed? How many more breasts must hit the trash before we know for sure if a virus causes breast cancer in women? How many more arms must be punctured and infused with toxic chemotherapy? How many more breasts must be irradiated, or when absent, left as barren scars, or artificially reconstructed? How many more children must lose their mothers, fathers their daughters, husbands their wives, and friends their friends, before we know the truth about this virus? The bodies that have succumbed to this disease are being buried at the rate of one per minute. The parts of women who are newly diagnosed with breast cancer are piling up at the rate of another pound of flesh every twenty seconds while we wait for answers. Billions of dollars are being hemorrhaged on a race for a cure that might be called off if we can complete this research. And who knows what other tumors this virus is capable of inducing: lung, lymphoma, endometrial are just a few of the latest suspects.

The end of breast cancer will be found in its beginning. If a virus is responsible for even a small portion of breast cancer, there are seven billion people on the planet who need to know about it as soon as possible. We can't afford to wait or waste another hundred years.

Afterword

Discovering a new tumor virus is a major event: a milestone on the leading edge of a rising curve of cases. The latest entry in this lineup is an interesting case in point. The bovine leukemia virus (BLV), which causes leukemia and lymphoma in cattle, and is also found in dairy cows, is notably ubiquitous. It infects 38 percent of beef herds, 85 percent of dairy cows, and 100 percent of dairy cows employed in large industrial operations. The addition of BLV to the list of animal tumor viruses found in humans is likely the biggest stone ever dropped in the pond of probable causes: its ripples feel more like waves. (Buehring G. C., "Exposure to Bovine Leukemia Virus Is Associated with Breast Cancer: A Case-Control Study," *PLOS*, September 2, 2015. See discussion below.)

BLV, which is closely related to the human T-cell leukemia virus (it's endemic in Japan and is shed in human breast milk), is the most common tumor virus in cattle and is carried in their blood. Despite its prevalence, BLV infection is not a major problem in the cattle industry. Of the 38 percent of animals infected, fewer than 5 percent will go on to develop cancer. When they do, they must be excluded from the herd, by law. One might ask how, and at what point, cattle ranchers know that their animals are infected with BLV. One might also ask how long cattle can be infected with BLV before they become sick or otherwise symptomatic. One might wonder how often healthy cattle are tested to see if they're infected with BLV, how this is done, and how reliable the tests are for detecting BLV infection. Even the unworried might be concerned that, on occasion—who knows how often—infected cattle may slip the knot between infection and detection, and pass the virus on to humans through their meat. Agricultural veterinarians can likely address all but the last question, which epidemiologists may one day be called upon to answer.

Pasteurization kills BLV found in the milk of dairy cows, but this decontamination process is not required in the cattle industry: raw and undercooked beef contains live virus. What's more, BLV targets breast tissue in precisely the same way that MTV[1] does—like a heat-seeking missile. Is this why meat-eating countries consistently have much higher rates of breast cancer? Read on.

The presence of BLV in cows may be a much larger problem for humans than the presence of BLV in beef. Like mice infected with MTV, cows infected with BLV shed the virus in their milk. The bridge to humans is not a stretch. The vast majority of people living in Western countries consume large amounts of cow's milk and cow's milk products during their lifetime, a lifetime that now extends, on average, nearly eighty years. Western countries, their larders jammed with dairy, are also regions where the highest rates of breast cancer in the world are found. Cold comfort that pasteurization kills the virus. The United States didn't mandate pasteurization until 1925, thousands of years after BLV entered the human population as a result of the frequent ingestion of raw milk and cheese. Today, unpasteurized cheese is still consumed, and preferred, in many parts of Europe. And although studies have never found that the consumption of raw milk and cheese increases the risk of leukemia in humans, once BLV enters the population, it has the potential to move from person to person just like any other virus.

In summary, the road between BLV and humans is paved with infected beef and dairy. Once BLV has crossed the border into humans, it might spread effectively from person to person. We don't know how.

Scientists at the University of California, Berkeley, decided to investigate BLV as a possible cause of human breast cancer. They tested samples of breast tissue—benign, premalignant, and malignant—obtained from 239 women who'd agreed to participate in a nationwide registry by donating samples of their breast tissue to the National Cancer Institute. The investigators found BLV in 59 percent of women with breast cancer, in 38 percent of women with premalignant breast disease, and in 29 percent of perfectly healthy women (who'd donated extraneous breast tissue that surgeons had removed during elective breast reduction).

The results of this study, published in September 2015, demonstrated (well beyond the possibility of chance) that there is a positive correlation between BLV infection and breast cancer. Additionally, the study revealed a positive gradient linking BLV infection to abnormal changes in the breast that are known to precede malignancy (i.e., atypia and ductal carcinoma in situ). The authors concluded that BLV infection is as strong a risk factor for breast cancer as other well-established factors: hormone use, reproductive history (i.e., early menarche, late menopause), diet, exercise, and alcohol consumption. Indeed, they found that only age and a positive family history of breast cancer were stronger risk factors than BLV infection. These same scientists reported in another, related study that in 6 percent of 219 volunteers, BLV appeared to be actively replicating in the blood (Buehring G. C., "Bovine Leukemia Virus DNA in Human Breast Tissue," *Emerging Infectious Diseases*, vol. 20, no. 5 [2014]).

As with MTV, the discovery that BLV might play a role in causing a large portion of human breast cancer needs to be independently confirmed via similar studies carried out at other institutions. Replication and convergence of these data are what's needed now to iron out the errors and hone the truth. But it's not hard to spot a rising tide when waters deepen and come farther into shore. Taken together, BLV and MTV may explain the so far inexplicable—why so many women around the world are getting breast cancer. Indeed, the rising incidence of breast cancer in nations that are modernizing and moving toward a Western diet may, in part, be the inadvertent consequence of the importation of milk and meat from countries where BLV is endemic—rampant, even—in animal herds.

The results of the most recent study from the University of California showcase yet another animal tumor virus, BLV, that might play a role in human cancer. The news ought to intensify our interest. Indeed, it ought to fix our attention there.

Of course, the news about BLV frightens as much as it impresses. But the goal is not to instill fear, but rather to inspire more good research. The contribution made by the scientists in Berkeley was enabled by the good will of generous, heroic women in Alabama, Pennsylvania, Ohio, and California who donated breast specimens in the name of science. Their generosity was multiplied by the enlightened effort of intrepid investigators willing to push against "the cure" to find the cause.

The bridge between the causes of breast cancer and its eradication will not be built out of cures, but rather by understanding and prevention. Inevitably, we'll look back and ask why it took so long. But there's enough data on the table now to create a mission: the time has come to puzzle out the viral connection in breast cancer, piece by piece.

Endnotes

Prologue: Lisa

1. No one seems to know exactly when and exactly what was said. It was likely first attributed to Fields in the December 5, 1925, *Buffalo Evening News*, "If at first you don't succeed, quit. Quit at once," by Stephen Leacock, page 6, Buffalo, New York.
2. There are two kinds of prevention, primary and secondary. Primary prevention *prevents the disease*. Secondary prevention *prevents the disease from recurring*. Komen has retooled the definition so that they market mammograms as preventing breast cancer when, in fact, they do no such thing. I use "The Pure Cure" to emphasize the importance of the true definition of primary prevention of breast cancer.

Chapter 2: Mendel

1. G. Mendel. "Experiments in plant hybridization," *Journal of the Royal Horticultural Society*, vol. 26 (1865): pp. 1–32.
2. Ibid.

Chapter 3: Lathrop and Loeb

1. A. Lathrop and L. Loeb. "Further investigations on the origins of tumors in mice," *Journal of Experimental Medicine*, vol. 22, no. 6 (December 1, 1915): pp. 713–31.

Chapter 4: Little and Jackson

1. K. Rader. *Making Mice: Standardizing Animals for American Biomedical Research, 1900–1955* (Princeton, N.J.: Princeton University Press, 2004), p. 83.
2. "Jobless Little," *Time*, vol. 7, no. 5 (February 4, 1929), p. 36.
3. Ibid.
4. Ibid.

Chapter 5: Bittner

1. A. Lathrop and L. Loeb, "Further investigations on the origin of tumors in mice: V. The tumor rate in hybrid strains," *Journal of Experimental Medicine*, vol. 28 (1918): pp. 457–500.
2. Staff of the Jackson Laboratory, "The existence of non-chromosomal influence in the incidence of mammary tumors in mice," *Science*, vol. 18 (November 1933): pp. 465–66.
3. J. Bittner. "Some possible effects of nursing on the mammary tumor incidence in mice," *Science*, vol. 84 (1936): p. 162.
4. J. Bittner. "Mammary tumors in mice in relation to nursing," *American Journal of Cancer*, vol. 30 (1937): pp. 530–38.
5. Sigmund Freud. Preface to the third (revised) edition, written by Freud in Vienna on March 15, 1931. This preface never appeared in a German edition of the work. *The Interpretation of Dreams* (1900) (New York: Basic Books, 2010): p. xxxii.
6. "Fundamental cancer research: Report of a committee appointed by the surgeon general," *Public Health Reports*, vol. 53 (1938): pp. 2121–30.
7. 7. M. Shimkin. "As memory serves: An internal history of the National Cancer Institute, 1937–57," *Journal of the National Cancer Institute*, vol. 59, no. 2, supplement (August 1977).
8. 8. G. Klein. *The Atheist and the Holy City: Encounters and Reflections*, translated by Theodore and Ingrid Freidman (Cambridge, MA.: MIT Press, 1990): p. 81.
9. "Fundamental cancer research: Report of a committee appointed by the surgeon general."
10. J. Murphy. "An analysis of trends in cancer research," *Journal of the American Medical Association*, vol. 1205 (1942): p. 107.

11. K. Rader. *Making Mice: Standardizing Animals for American Biomedical Research, 1900–1955* (Princeton, N.J.: Princeton University Press, 2004): p. 115.

Chapter 6: McKnight and Christian

1. Minneapolis *Journal*, Sunday, December 30, 1906.
2. *Handbook of Private Schools*, Porter E. Sargent author and publisher, Boston, p. 620 (1918).
3. Ibid.
4. *Harvard College Class of 1895 Second Report*, Harvard College, University Press (1902), p. 31.

Chapter 7: Andervont

1. M. Shimkin. "As memory serves," *Journal of the National Cancer Institute*, vol. 59, no. 2, supplement (August 1977): p. 577.
2. Ibid.
3. Ibid.
4. A. Andervont and W. R. Bryan. *A Symposium on Mammary Tumors in Mice* (1945): pp. 123–39.

Chapter 8: Gross, Huebner, and Baker

1. In a paper he published in 1985, *New York Times*, July 22, 1999, "Ludwik Gross, a trailblazer in cancer research, dies at 94," by Lawrence K. Altman.
2. Ibid.
3. *Robert Huebner, A Virologist's Odyssey*, by Edward A. Berman, MD, Office of NIH History, 2005, p. 36.
4. Carl Baker, MD, recalled in *An Administrative History of the National Cancer Institute's Viruses and Cancer Program, 1950–1972*, Carl G. Baker, MD, National Institutes of Health Office of History (2004), p. 9.
5. M. Matsuoka. "Human T-cell leukemia virus type-1 (HTLV-1) infectivity," *Nature Reviews Cancer*, vol. 7 (April 2007): pp. 270–80.
6. Congressional Appropriations Committee in 1958. *An Administrative History of the National Cancer Institute's Viruses and Cancer Program, 1950–1972*, Carl G. Baker, MD, National Institutes of Health Office of History (2004), p. 13.

7. Baker, *An Administrative History of the National Cancer Institute's Viruses and Cancer Program, 1950–1972*, p. 13.
8. Ibid.
9. Ibid.
10. Ibid.
11. Ibid.
12. Ibid.
13. Ibid.
14. Ibid.
15. Ibid.
16. Ibid.
17. Ibid.
18. Ibid.
19. Ibid.
20. Ibid.

Chapter 9: Mary Lasker

1. The quotes in this chapter, unless otherwise specified, are all taken from the oral history Mary Lasker recorded for Columbia University's Notable New Yorkers series, 1962–82.
2. R. A. Rettig. *Cancer Crusade: The Story of the National Cancer Act of 1971* (Authors Choice Press, 1997): p. 41.
3. Ibid.
4. Ibid.
5. "Cancer Army," *Time,* vol. 29, no. 12 (March 22, 1937): p. 56. No author credit given.
6. Ibid.
7. Ibid.
8. *Early Detection: Women, Cancer, and Awareness Campaigns in the Twentieth-Century United States*, by Kirsten Elizabeth Gardner, University of North Carolina Press, 2006, p. 102.
9. Bernadine Bailey, "An ounce of prevention: Today's cure for cancer," *Reader's Digest* (October 1944): p. 102.
10. Ibid.

11. Paul de Kruif. "Fifty thousand could live," *Reader's Digest* (November 1944): p. 89.
12. Ibid.
13. R. A. Rettig. *Cancer Crusade*, p. 41.
14. Ibid.

Chapter 10: Garb

1. *Out of Africa*, directed by Sidney Pollack (1985). Screenplay by Kurt Luedtke. Based on *Out of Africa* and other writings by Karen Blixen, *Isak Dinesen: The Life of a Storyteller* by Judith Thurman, and *Silence Will Speak* by Errol Trzebinski.
2. Mary Lasker, in an interview for Notable New Yorkers, Columbia University Libraries Oral History Research Office. Available at http://www.columbia.edu.
3. Ibid.
4. Ibid.
5. Ibid.
6. Ibid.
7. Robert Hickey. *Honor and Respect, second edition* (Washington, DC: Protocol School of Washington).
8. S. Garb. *Cure for Cancer: A National Goal* (New York: Springer, 1968): p. 83.
9. Ibid.
10. Ibid.
11. Ibid.
12. Ibid.
13. Ibid.
14. W. Henle. "Epidemiologic evidence for causal relationship between Epstein-Barr virus and Burkitt's lymphoma from Ugandan prospective study," *Nature*, vol. 274, no. 5873 (August 24, 1978): pp. 756–61.
15. Ibid.

Chapter 11: The Congress of the United States

1. Mary Lasker, in an interview for Notable New Yorkers, Columbia University Libraries Oral History Research Office. Available at http://www.columbia.edu.
2. Ibid.

3. Ibid.
4. Ibid.
5. Ibid.
6. Ibid.
7. Ibid.
8. Ibid.
9. Ibid.
10. Salvador Luria, MD, a Nobel laureate, and a professor at the Massachusetts Institute of Technology; *Congressional Record*, 92nd Congress, 1st session, 1971, p. 117.
11. Lasker, Notable New Yorkers.
12. Ibid.
13. Ibid.

Chapter 12: Zinder and zur Hausen

1. Carl Baker, MD. *An Administrative History of the National Cancer Institute's Viruses and Cancer Program, 1950–1972.* Available at http://history.nih.gov.
2. National Institutes of Health, Office of History, interview with Dr. Robert Gallo by Dr. Victoria Harden, August 25, 1994, p. 28.
3. Samuel Herman. Available at http://history.nih.gov/archives/downloads/samuelherman.pdf.
4. Mary Lasker, in an interview for Notable New Yorkers, Columbia University Libraries Oral History Research Office. Available at http://www.columbia.edu.
5. Harold zur Hausen. "The search for infectious causes of human cancers: Where and why," Nobel Lecture (2008). Available at http://nobelprize.org. (Also, *Finding the Viral Link: The story of Harald zur Hausen,*" *Cancer World*, by Peter McIntyre, July–August, 2005, p. 36.)
6. Ibid.
7. Ibid.

Chapter 13: Moore and Ogawa

1. "A new virus in a spontaneous mammary tumor of a rhesus monkey," H. Chopra, M. Mason, *Cancer Research*, vol. 30 (August 1970): p. 2081. D. Moore. "Search for a human breast cancer virus," *Nature*, vol. 229, no. 5287 (1971): pp. 611–14.

2. Ibid.
3. Ibid.
4. R. Axel. "Presence in human breast cancer of RNA homologous to mouse mammary tumor virus RNA," *Nature*, vol. 235 (1971): pp. 32–36.
5. S. Zotter. "Mouse mammary tumor virus-related antigen in core-like density fractions from large samples of women's milk," *European Journal of Cancer*, vol. 16 (1980): pp. 455–67.
6. H. Ogawa. "Occurrence of antibodies against intracytoplasmic A-particles of mouse mammary tumor virus in sera of breast cancer patients," *Gan*, vol. 69, no. 4 (1978): pp. 539–44.
7. Ibid.
8. Harold Varmus, MD. Harvard University Commencement Address (June 6, 1996). Available at http://harvardmagazine.com.
9. Ibid.
10. Ibid.
11. M. Kumar. *Quantum* (Cambridge, U.K.: Icon Books, 2008): p. 275.

Chapter 14: Baltimore, Temin, Nusse, Callahan, and Franklin

1. G. C. Franklin. "Expression of human sequences related to those of mouse mammary tumor virus," *Journal of Virology*, vol. 62 (1988): pp. 1203–10.
2. Ibid.
3. Ibid.
4. I. Evans, "Challenge of breast cancer," Lancet Conference, April 30, 1994, 343, 1085-6
1. Ibid.

Chapter 15: Evans, Mueller, and Stewart

2. T. Stewart. "Breast cancer incidence highest in the range of one species of house mouse, *Mus domesticus*," *British Journal of Cancer*, vol. 82 (2000): pp. 446–51.

Chapter 16: Pogo, Holland, and Etkind

1. Y. Wang and B. Pogo. "Detection of mouse mammary tumor virus envelope gene-like sequences in human breast cancer," *Cancer Research*, vol. 55 (1995): pp. 5173–79.
2. Ibid.

3. J. Holland. "Human mammary tumor virus." *Breast Cancer Research Treatment*, vol. 100 (2006): abstract 6.
4. Ibid.
5. P. Etkind. "Clonal isolation of different strains of mouse mammary tumor virus-like DNA sequences from both the breast tumors and non-Hodgkin's lymphomas of individual patients diagnosed with both malignancies," *Clinical Cancer Research*, vol. 10 (2004): pp. 5656–64.

Chapter 17: Lawson, Ford, Garry, and Levine

1. J. Lawson. "From Bittner to Barr: A viral, diet and hormone breast cancer aetiology hypothesis," *British Cancer Research*, vol. 3 (2001): pp. 81–85.
2. Ibid.
3. C. Ford. "Progression from normal breast pathology to breast cancer is associated with increasing prevalence of mouse mammary tumor virus-like sequences in men and women," *Cancer Research*, vol. 64 (July 15, 2004): p. 4755.
4. Ibid.
5. Ibid.
6. R. Garry. "Of mice, cats, and men: Is human breast cancer a zoonosis?" *Microscopic Research Technology*, vol. 68, nos. 3–4 (November 2005): pp. 197–208.
7. Ibid.
8. P. Levine. "Immunopathologic features of rapidly progressing breast cancer in Tunisia," *Proceedings American Association Cancer Research*, vol. 21 (1980): p. 170.
9. P. Levine. "Human mammary tumor virus in inflammatory breast cancer," *Cancer*, vol. 116, supplement 11 (June 1, 2010): pp. 2741–44.

Chapter 18: Pascrell, Niederhuber, Callahan, Rouault, and Arbus

1. F. Rouault. "Rapid spread of mouse mammary tumor virus in cultured human breast cells," *Retrovirology*, vol. 47 (2007): p. 73.
2. Ibid.
3. "Nikki Weiss and Dr. Susan Love, MD, speak out about beating breast cancer," Victoria Brownworth, http://www.curvemag.com/Curve-Magazine/Web-Articles-2013/We-Can.

Chapter 19: The National Institutes of Health

1. J. Holland. "The human mammary tumor virus," Clinical Center Grand Rounds of the National Institutes of Health (July 10, 2013). Available at http://videocast.nih.gov.
2. A reference to a videotape of Holland's presentation at the National Institutes of Health in July 2013 is given in the bibliography at the back of the book. I would encourage you to watch it. Even for the nonscientist, it is easy to understand and powerful.

Afterword

1. A word about the naming of the "breast cancer virus" would be in order, especially now that converging and compelling data increasingly point to it as a cause of human breast cancer. When John Bittner discovered that a virus caused breast cancer in mice he, naturally, considered it to be a "breast cancer virus." When he subsequently discovered that it also caused lymphoma, especially in male mice, he did not change the name of the virus nor did anyone else for many years. The virus became known as the "mouse mammary tumor virus" (MMTV). But later, when other scientists found essentially the same virus in cats, this variant of the virus was given the name "feline mammary tumor virus." When it was discovered in dogs, it was given the name "canine mammary tumor virus." When Beatriz Pogo of Mount Sinai found the virus in human breast cancer tissue, she named it "human mammary tumor virus" (HMTV). Since the virus is essentially the same virus, adapted in many species, the more recent convention is simply to refer to it as the "mammary tumor virus" (MTV).

Bibliography

Almeida, J. "Virus-like particles in blood of two acute leukemia patients," *Science*, vol. 142 (1963): p. 1487.

Andersson, A. "Expression of human endogenous retrovirus ERV3 (HERV-R) mRNA in normal and neoplastic tissues," *International Journal of Oncology*, vol. 12 (1998): pp. 309–13.

Andervont, H. "Problems concerning the tumor viruses," in F. M. Burnett and W. M. Stanley, editors, *The Viruses* (New York: Academy Press, 1959).

Andervont, H., and W. Bryan. "Properties of the mouse mammary-tumor agent," *Journal of the National Cancer Institute*, vol. 5 (1944): p. 143.

Axel, R. "Presence in human breast cancer of RNA homologous to mouse mammary tumour virus RNA," *Nature*, vol. 235 (1972): pp. 32–36.

Bang, F. "Electron microscopic evidence concerning the mammary tumor inciter (Virus). I. A study of normal and malignant cells from the mammary gland of mice." *Bulletin of Johns Hopkins Hospital*, vol. 98 (1956): p. 255.

Bang, F. "Electron microscopic evidence concerning the mammary tumor inciter (virus). II. An electron microscopic study of spontaneous and induced mammary tumors of mice," *Bulletin of Johns Hopkins Hospital*, vol. 98 (1956): p. 287.

Bell, T. "Isolation of a retrovirus from a case of Burkitt's lymphoma," *British Medical Journal*, vol. 1 (1964): p. 1212.

Benit, L. "Identification, Phylogeny, and Evolution of Retroviral Elements Based on Their Envelope Genes," *Journal of Virology*, vol. 75, no. 23 (2001): pp. 11709–19.

Bera, T. "Defective retrovirus insertion activates c-Ha-ras proto-oncogene in an MNU-induced rat mammary carcinoma," *Biochemical and Biophysical Research Communications*, vol. 248 (1998): pp. 835–40.

Berkhout, B. "Identification of an active reverse transcriptase enzyme encoded by a human endogenous HERV-K retrovirus," *Journal of Virology*, vol. 73, no. 3 (1999): pp. 2365–75.

Bindra, A. "Search for DNA of exogenous mouse mammary tumor virus-related virus in human breast cancer samples," *Journal of General Virology*, vol. 88 (2007): 1806–9.

Bittner, J. "Some possible effects of nursing on the mammary gland tumor incidence in mice," *Science*, vol. 84 (August 14, 1936): p. 162.

Bittner, J. "The influence of transplanted normal tissue on breast cancer ratios in mice," *Public Health Report*, vol. 54 (1939): p. 1827.

Bittner, J. "The milk-influence of breast tumors in mice," *Science*, vol. 95 (May 1942): pp. 462–63.

Blair, P. "Neutralization of the mouse mammary tumor virus by rabbit antisera against C3Hf tissue," *Cancer Research*, vol. 23 (1963): p. 381.

Bock, M. "Endogenous retroviruses and the human germline," *Current Opinions in Genetics and Development*, vol. 10 (2000): pp. 651–55.

Burger D. "Virus-like particles in human leukemic plasma," *Proceedings of the Society for Experimental Biology and Medicine*, vol. 115 (1964): p. 151.

Burkitt, D. "A lymphoma syndrome in African children," *Annals of the Royal College of Surgeons*, vol. 30 (1962): p. 211.

Burkitt D. "Geographical and tribal distribution of the African Lymphoma in Uganda," *British Medical Journal*, vol. 1 (1966): p. 569.

Callahan, R. "Detection and cloning of human DNA sequences related to the mouse mammary tumor virus genome," *Proceedings of the National Academy of Science USA*, vol. 79 (1982): pp. 5503–7.

"Cancer Virus," *Time* (March 18, 1946).

Dalton A. "Some ultrastructural characteristics of a series of primary and transplanted plasma-cell tumors of the mouse," *Journal of the National Cancer Institute*, vol. 26, no. 5 (1961): pp. 1221–66.

Day N. "Antibodies reactive with murine mammary tumor virus in sera of patients with breast cancer: Geographic and family studies," *Proceedings of the National Academy of Science USA*, vol. 78 (1984): pp. 2483–87.

DeOme, K. "The mouse mammary tumor virus," *Fed Proc*, vol. 12 (1962): p. 15.

Dion, A. "A human protein related to the major envelope protein of murine mammary tumor virus: identification and characterization," *Proceedings of the National Academy of Science USA*, vol. 77 (1980): pp. 1301–5.

Dion, A. "Retrovirus association with breast cancer: a critical appraisal," *Breast Cancer Research and Treatment*, vol. 9 (1987): pp. 155–56.

Dmochowski, L. "The milk agent in the origin of mammary tumors in mice," *Advances in Cancer Research*, vol. 1 (1953): p. 103.

Ekbom, A. "Breast-feeding and breast cancer in the offspring," *British Journal of Cancer*, vol. 67 (1993): pp. 2375–82.

Epstein, M. "Morphological and biological studies on a virus in cultured lymphoblasts from Burkitt's lymphoma," *Journal of Experimental Medicine*, vol. 121 (1965): p. 761.

Etkind, P. "Clonal isolation of different strains of mouse mammary tumor virus-like DNA sequences from both the breast tumors and non-Hodgkin's lymphomas of individual patients diagnosed with both malignancies," *Clinical Cancer Research*, vol. 10 (2004): pp. 5656–64.

Etkind, P. "Mouse mammary tumor virus-like ENV gene sequences in human breast tumors and in a lymphoma of a breast cancer patient," *Clinical Cancer Research*, vol. 6 (2000): pp. 1273–78.

Faedo, M. "Mouse mammary tumor-like virus is associated with p53 nuclear accumulation and progesterone receptor positivity but not estrogen positivity in human female breast cancer," *Clinical Cancer Research*, vol. 10 (2004): pp. 4417–19.

Fennelly, J. "Co-amplification of tail-to-tail copies of MuRVY and APE retroviral genomes on the Mus musculus Y Chromosome,"*Mammalian Genome*, vol. 7 (1996): pp. 31–6.

Fernandez-Cobo, M. "Transcription profiles of non-immortalized breast cancer cells," *Biomed Central Cancer*, vol. 6 (2006): pp. 99–100.

Ford, C. "MMTV virus-like RNA transcripts and DNA are found in affected cells of human breast cancer," *Clinical Cancer Research*, vol. 10 (2004): pp. 7284–89.

Ford, C. "Mouse mammary tumor virus-like gene sequences in breast tumors of Australian and Vietnamese women," *Clinical Cancer Research*, vol. 9 (2003): pp. 1118–20.

Franklin, G. "Expression of human sequences related to those of mouse mammary tumor virus," *Journal of Virology*, vol. 62 (1988): pp. 1203–10.

Friedman, G. "Spousal concordance for cancer incidence: A cohort study," *Cancer*, vol. 86 (2000): pp. 2413–19.

Fukuoka, H. "No association of mouse mammary tumor virus-related retrovirus with Japanese cases of breast cancer," *Journal of Medical Virology*, vol. 80 (2008): pp. 1447–51.

Garb, S. *Cure for Cancer: A National Goal* (New York: Springer Publishing, 1968).

Gay, F. "Morphogenesis of Bittner Virus," *Journal of Virology*, vol. 5 (June 1970): pp. 801–16.

Glenn, W. "Presence of MMTV-like gene sequences may be associated with specific breast cancer morphology," *Journal of Clinical Pathology*, vol. 60, no. 9 (2007): p. 1071.

Golovkina T. "Coexpression of exogenous and endogenous mouse mammary tumor virus RNA in vivo results in viral recombination and broadens the virus host range," *Journal of Virology*, vol. 68 (1994): pp. 5019–26.

Golovkina, T. "A novel membrane protein is a mouse mammary tumor virus receptor," *Journal of Virology*, vol. 72 (1998): pp. 3066–71.

Golovkina, T. "Superantigen activity is needed for mouse mammary tumor virus spread within the mammary gland," *Journal of Immunology*, vol. 161 (1998): pp. 2375–82.

Gotlieb-Stematsky, T. "Increased tumor formation by polyoma virus in the presence of non-oncogenic viruses," *Nature*, vol. 212 (1966): p. 421.

Gray, D. "Activation of int-1 and int-2 loci in GRf mammary tumors," *Virology*, vol. 154 (1986): pp. 271–278.

Griffiths, D. "Endogenous retroviruses in the human genome sequence," *Genome Biology*, vol. 2, no. 6 (2001): pp. 1017.1–1017.5.

Hachana, M. "Prevalence and characteristics of the MMTV-like associated breast carcinomas in Tunisia," *Cancer Letter*, vol. 271 (2008): pp. 222–30.

Hareuveni, M. "Breast cancer sequences identified by mouse mammary tumor (MMTV) antiserum are unrelated to MMTV," *International Journal of Cancer*, vol. 46 (1990): pp. 1134–35.

Held, W. "Reverse transcriptase-dependent and -independent phases of infection with mouse mammary tumor virus: Implications for superantigen function," *Journal of Experimental Medicine*, vol. 180 (1994): pp. 2347–51.

Heller, J. "Research on cancer viruses," *Public Health Report*, vol. 75 (1960): p. 501.

Herbut, P. "Human leukemia virus in mice," *Arch Path*, vol. 83 (1967): p. 123.

Holland, J. *Breast Cancer Research Treatment*, vol. 11 (2006): abstract 6.

Holland, J. "The human mammary tumor virus," National Institutes of Health Clinical Center Grand Rounds (July 10, 1013). Available at http://videocast.nih.gov.

Howard, D. "Isolation of a series of novel variants of murine mammary tumor viruses with broadened host range," *International Journal of Cancer*, vol. 25 (1980): pp. 647–54.

Hughes, J. "Evidence for genomic rearrangements mediated by human endogenous retroviruses during primate evolution," *Nature Genetics*, vol. 29 (2001): pp. 487–89.

"Identity crisis," *Nature*, vol. 457 (2009): pp. 935–36.

Imai, S. "Distribution of mouse mammary tumor virus in Asian wild mice," *Journal of Virology*, vol. 68 (1994): pp. 3437–42.

Indik, S. "Mouse mammary tumor virus infects human cells," *Cancer Research*, vol. 65 (2005): pp. 6651–59.

Indik, S. "A novel, mouse mammary tumor virus encoded protein with Rev-like properties," *Virology*, vol. 337 (2005): pp. 1–6.

Indik, S. "Rapid spread of mouse mammary tumor virus in cultured human breast cells," *Retrovirology*, vol. 4 (2007): p. 73.

Karlsson, H. "Retroviral RNA identified in the cerebrospinal fluids of brains of individuals with schizophrenia," *PNAS*, vol. 98, no. 6 (2001): pp. 4634–39.

Katz, E. "MMTV env encodes an ITAM responsible for transformation of mammary epithelial cells in three-dimensional culture," *Journal of Experimental Medicine*, vol. 201 (2005): pp. 431–39.

Kerin, M. "Synchronous metastatic breast cancer in husband and wife," *Irish Journal of Medical Science*, vol. 165 (1996): p. 50.

Kidd, J. "The enduring partnership of a neoplastic virus and carcinoma cells: Continued increase of virus in the V2 carcinoma during propagation in virus immune hosts," *Journal of Experimental Medicine*, vol. 75 (1942): p. 7.

Kuhn, T. *The Structure of Scientific Revolutions, third edition.* (Chicago, IL: University of Chicago Press, 1996).

Larson, E. "Human endogenous proviruses," *Current Topics in Microbiology and Immunology*, vol. 148 (1989): pp. 115–32.

Lawson J. "From Bittner to Barr: a viral, diet and hormone breast cancer aetiology hypothesis," *Breast Cancer Research*, vol. 3 (2001): 81–85.

Leib, C. "Endogenous retroviral elements in human DNA," *Cancer Research* (supplement), vol. 50 (1990): pp. 5636–42.

Levine, P. "Increased detection of breast cancer virus sequences in inflammatory breast cancer," *Advances in Tumor Virology.* vol. 1 (2009): pp. 3–7

Levine, P. "Increased incidence of mouse mammary tumor virus-related antigen in Tunisian patients with breast cancer," *International Journal of Cancer*, vol. 3 (1984): pp. 305–8.

Levine, P. "Increasing evidence for a human breast carcinoma virus with geographic differences," *Cancer*, vol. 101 (2004): pp. 721–26.

Links, J. "The growth accelerating effect of Bittner virus in monolayers of baby mouse kidney cells," *Journal of General Virology*, vol. 5 (1969): pp. 547–50.

Litvinov, S. "Expression of proteins immunologically related to murine mammary tumor virus (MMTV) core proteins in cells of breast cancer continuous cell lines MCF7, T47D, MDA-MB231 and cells from human milk," *Acta Virologica*, vol. 33 (1989): pp. 137–42.

Liu, B. "Identification of a proviral structure in human breast cancer," *Cancer Research*, vol. 61 (2001): pp. 1754–59.

Lloyd, R. "Murine mammary tumor virus related antigen in human male mammary carcinoma," *Cancer*, vol. 51 (1983): pp. 654–61.

Lower, R. "A general method for the identification of transcribed retrovirus sequences (R-U5 PCR) reveals the expression of the human endogenous retrovirus loci HERV-H and HERV-K in teratocarcinoma cells," *Virology*, vol. 192 (1992): pp. 501–11.

Lower, R. "The pathogenic potential of endogenous retroviruses: facts and fantasies," *Trends in Microbiology*, vol. 7, no. 9 (1999): pp. 350–56.

Luo, T. "Study of mouse mammary tumor virus-like gene sequences expressing breast tumors of Chinese women," *Sichuan Da Xue Xue Bao Yi Xue Ban*, vol. 37 (2006): pp. 844–46.

Lushnikova, A. "Detection of the env MMTV-homologous sequences in mammary carcinoma patient intestine lymphoid tissue," *Doklady Biological Sciences*, vol. 399 (2004): pp. 423–26.

Lynch, H. "Is cancer communicable?" *Medical Hypotheses*, vol. 14 (1984): pp. 181–98.

MacMahon, B. "Etiology of human breast cancer: A review," *Journal of the National Cancer Institute*, vol. 50 (1973): pp. 21–42.

Mager, D. "HERV-H endogenous retroviruses: presence in the new world branch but amplification in the old world primate lineage," *Virology*, vol. 213 (1995): pp. 395–404.

Mager, D. "Novel mouse type D endogenous proviruses and elements share long terminal repeat and internal sequences," *Journal of Virology*, vol. 74, no. 16 (2000): pp. 7221–29.

Mant, C. "A human murine mammary tumor virus-like agents are genetically distinct from endogenous retroviruses and not detectable in breast cancer cell lines of biopsies," *Virology*, vol. 318 (2004): pp. 393–403.

Marchetti, A. "Host genetic background effect on the frequency of mouse mammary tumor virus-induced rearrangements of the int-1 and int-2 loci in mouse mammary tumors," *Journal of Virology*, vol. 65, no. 8 (1991): pp. 4550–54.

Marrack, P. "A maternally inherited superantigen encoded by a mammary tumour virus," *Nature*, vol. 349 (1991): pp. 524–25.

Melana, S. "Characterization of viral particles isolated from primary cultures of human breast cancer cells," *Cancer Research*, vol. 67 (2007): pp. 8960–65.

Melana, S. "Search for mouse mammary tumor virus-like env sequences in cancer and normal breast from the same individuals," *Clinical Cancer Research*, vol. 7 (2001): pp. 283–84.

Mesa-Tejada, R. "Detection in human breast carcinomas of an antigen immunologically related to a group-specific antigen of mouse mammary tumor virus," *Proceedings of the National Academy of Science USA*, vol. 75 (1978): pp. 1529–33.

Mettenleiter, W. "Isolation of viral agents from human blood and their relationship to lymphatic leukemia," *Oncologia*, vol. 16 (1963): p. 307.

Moore, D. "Search for a human breast cancer virus," *Nature*, vol. 229 (1971): pp. 611–15.

Morton, D. "Acquired immunological tolerance to spontaneous mammary adenocarcinomas following neonatal infection with mammary tumor agent," *Proceedings of the American Association of Cancer Research*, vol. 5 (1964): p. 46.

Mueller-Lantzsch, N. "Human endogenous retroviral element K10 (KERV-K10) encodes a full-length gag homologous 73-kDa protein and a functional protease," *AIDS Research and Human Retroviruses*, vol. 9, no. 4 (1993): pp. 343–50.

Nusse, R. "Insertional mutagenesis in mouse mammary tumorigenesis," *Current Topics in Microbiology and Immunology*, vol. 171 (1991): pp. 43–65.

Nusse, R. "The int genes in mammary tumorigenesis and in normal development," *TIG*, vol. 4, no. 10 (1988): pp. 291–95.

Ono, M. "Stimulation of expression of the human endogenous retrovirus genome by female steroid hormones in human breast cancer cell line T47D," *Journal of Virology*, vol. 61 (1987): pp. 2059–62.

Paces, J. "HERV: database of human endogenous retroviruses," *Nucleic Acids Research*, vol. 30, no. 1 (2002): pp. 205–6.

Palmarini, M. "The exogenous form of jaagsiekte retrovirus is specifically associated with a contagious lung cancer of sheep," *Journal of Virology*, vol. 70, no. 3 (1996): pp. 1618–23.

Peters, G. "Tumorigenesis by mouse mammary tumor virus: Evidence for a common region for provirus integration in mammary tumors," *Cell*, vol. 33, no. 2 (1983): 369–77.

Pitelka, D. "Virus-like particles in precancerous hyperplastic mammary tissues of C3H and C3hf mice," *Journal of the National Cancer Institute*, vol. 25 (1960): p. 753.

Pogo, B. "Detection of mammary tumor virus env gene-like sequences in human breast cancer," *Cancer Research*, vol. 55 (1995): pp. 5173–79.

Pogo, B. "Possibilities of a viral etiology for human breast cancer: A review," *Biological Trace Element Research*, vol. 56 (1997): pp. 131–42.

Pogo, B. "Sequences homologous to the MMTV env gene in human breast carcinoma correlate with overexpression of laminin receptor," *Clinical Cancer Research*, vol. 5 (1999): pp. 2108–11.

Prak, E. "Mobile elements and the human genome," *Nature Reviews Genetics*, vol. 1 (2000): pp. 134–44.

Prince, A. "Virus like-particles in human plasma and serum from leukemic, hepatitic and control patients," *Fed Proc*, vol. 24 (1965): p. 175.

Rettig, Richard A. *Cancer Crusade: The Story of the National Cancer Act of 1971.* Authors Choice Press, New York, 2005 (Princeton University Press, 1977).

Reus, K. "HERV-K (OLD): Ancestor sequences of the human endogenous retrovirus family HERV-K (HNL-2)," *Journal of Virology*, vol. 75, no. 19 (2001): pp. 8917–26.

Reuss, F. "cDNA sequence and genomic characterization of intracisternal a-particle-related retroviral elements containing an envelope gene," *Journal of Virology*, vol. 65, no. 11 (1991): pp. 5702–9.

Rous, P. "A sarcoma of the fowl transmissible by an agent separable from the tumor cells," *Journal of Experimental Medicine*, vol. 13 (1911): p. 397.

Rous, P. "Transmission of a malignant new growth by means of a cel-free filtrate," *Journal of the American Medical Association*, vol. 56 (1911): p. 198.

Russ, J. "Identical cancers in husband and wife," *Surgery, Gynecology, and Obstetrics*, vol. 150 (1980): pp. 664–66.

Salmons, B. "Production of mouse mammary tumor virus upon transfection of a recombinant proviral DNA into cultured cells," *Virology*, vol. 144 (1985): pp. 101–14.

Sarkar, N. "Type B virus and human breast cancer," in *The Role of Viruses in Human Cancer, Volume 1*, A. Giraldo and A. Beth, editors (North Holland: Elsevier, 1980): pp. 207–35.

Shackleford, G. "Mouse mammary tumor virus infection accelerates mammary carcinogenesis in Wnt-1 transgenic mice by insertional activation of int-1/Fgf-3 and hst/Fgf-4," *Proceedings of the National Academy of Science USA*, vol. 90 (1993): pp. 740–44.

Smit, A. "Interspersed repeats and other mementos of transposable elements in mammalian genomes," *Current Opinion in Genetics & Development*, vol. 9 (1999): pp. 657–63.

Steintz, R. "Male breast cancer in Israel: selected epidemiological aspects," *Israel Journal of Medical Science*, vol. 17 (1981): pp. 816–21.

Stewart, A. "Identification of human homologues of the mouse mammary tumor virus receptor," *Archives of Virology*, vol. 147 (2002): pp. 577–81.

Stewart, S. "Burkitt's tumour: tissue culture, cytogenetic and virus studies," *Journal of the National Cancer Institute*, vol. 34 (1964): p. 319.

Stewart, T. "Breast cancer incidence highest in the range of one species of house mouse, *Mus domesticus*," *British Journal of Cancer*, vol. 82 (2000): pp. 446–51.

Sumidaie, A. "Particles with properties of retroviruses in monocytes from patients with breast cancer," *Lancet*, vol. 1, nos. 8575–76 (1988): pp. 5–9.

Szabo, S. "Of mice, cats and men: Is human breast cancer a zoonosis?" *Microscopy Research and Technique*, vol. 68 (2005): pp. 197–208.

Tchenio, T. "Defective retroviruses can disperse in the human genome by intracellular transposition," *Journal of Virology*, vol. 65 (1991): pp. 2113–18

Tonjes, R. "Characterization of human endogenous retrovirus type K virus-like particles generated from recombinant baculoviruses," *Virology*, vol. 233 (1997): pp. 280–91.

Tristem, M. "Identification and characterization of novel human endogenous retrovirus families by pylogenetic screening of the human genome mapping project database," *Journal of Virology*, vol. 74, no. 8 (2000): pp. 3715–30.

Tsubura, A. "Intervention of T-cells in transportation of mouse mammary tumor virus to mammary gland cells in vivo," *Cancer Research*, vol. 48 (1988): pp. 6555–59.

Turner, G. "Insertional polymorphisms of full-length endogenous retroviruses in humans," *Current Biology*, vol. 11 (2001): pp. 1531–35.

Wang, Y. "Detection of mammary tumor virus env gene-like sequences in human breast cancer," *Cancer Research*, vol. 55 (1995): pp. 5173–79.

Wang, Y. "Detection of MMTVlike LTR and LTR-env gene sequences in human breast cancer," *International Journal of Oncology*, vol. 18 (2001): pp. 1041–44.

Wang, Y. "Expression of mouse mammary tumor virus-like env gene sequences in human breast cancer," *Clinical Cancer Research*, vol. 4 (1998): pp. 2565–68.

Wang, Y. "A mouse mammary tumor virus-like long terminal repeat superantigen in human breast cancer," *Cancer Research*, vol. 64 (2004): pp. 4105–11.

Wang, Y. "Presence of MMTV-like env gene sequences in gestational breast cancer," *Medical Oncology*, vol. 20 (2003): pp. 233–36.

Wetchler, B. "Carcinoma of the breast occurring in a husband and wife: a brief communication," *Mount Sinai Journal of Medicine*, vol. 42 (1975): pp. 205–6.

Westley, B. "The human genome contains multiple sequences of varying homology to mouse mammary tumour virus DNA," *Gene*, vol. 28 (1984): pp. 221–27.

Whitaker. "MMTV-induced mammary tumorgenesis: gene discovery, progression to malignancy and cellular pathways," *Oncogene*, vol. 19 (2006): pp. 992-1001.

Witkin, A. "Antigens and antibodies cross-reactive to the murine mammary tumor virus in human breast cyst fluids," *Journal of Clinical Investigation*, vol. 67 (1981): pp. 216–22.

Witt, A. "The mouse mammary tumor virus-like sequence is not detectable in breast cancer tissue of Austrian patients," *Oncology Report*, vol. 10 (2003): pp. 1025–29.

Wright, D. "Burkitt's tumor and childhood lymphosarcoma," *Clinical Pediatrics*, vol. 6 (1967): p. 116.

Xu, L. "Does a betaretrovirus infection trigger primary biliary cirrhosis?" *Proceedings of the National Academy of Sciences*, vol. 100 (2003): pp. 8454–59.

Yang, M. "Presence of a mouse mammary tumor virus MMTV-related antigen in human breast carcinoma cells and its absence from normal epithelial cells," *Journal of the National Cancer Institute*, vol. 61 (1978): pp. 1205–7.

Zammarchi, F. "MMTV-like sequences in human breast cancer: a fluorescent PCR/laser microdissection approach," *Journal of Pathology*, vol. 209 (2006): pp. 436–44.

Zangen, R. "Mouse mammary tumor-like envgene as a molecular marker for breast cancer?" *International Journal of Cancer*, vol. 102 (2002): pp. 304–7.

Zapata-Benavides, P. "Mouse mammary tumor virus-like gene sequences in breast cancer samples of Mexican women," *Intervirology*, vol. 50 (2007): pp. 402–7.

Zotter, S. "Mouse mammary tumour virus-related antigens in core-like density fractions from large samples of women's milk," *European Journal of Cancer*, vol. 16 (1980): pp. 455–67.

Acknowledgments

I would like to thank medical librarian Arlene Mangino for her tireless support and encouragement over the past eight years. I am so grateful for every article, cheerfully tracked down, that you produced, often at a moment's notice. Thank you so very much. I recall my history professor in college, Josephine Pacheco, telling me in my freshman year at George Mason University, in 1973, that "librarians are the most wonderful people in the world." She was right!

In 1975, I left a wonderful job working as a medical secretary for Donald Fishman, MD, pediatric neurologist at the Children's Hospital in Washington, DC. I was off to George Mason University to study full time, and hopefully become a writer. Dr. Fishman gave me a lovely going away gift of bright yellow Crane stationary, a choice that I'm sure his charming wife, Renee, had selected. It included a note that said, "Good luck. Please send a signed copy of a first edition." I'll never forget those words of encouragement, which meant so much to me at the time and that have stayed with me for the past forty years. Sorry it took so long.

I'd like to thank the thousands of women in my practice who teach me every day how important this work is and how wonderful it is to be in this profession.

Index

Y

Z